VOLUME	EDITOR-IN-CHIEF	PAGES	
34	William S. Johnson	121	*Out of print*
35	T. L. Cairns	122	*Out of print*
36	N. J. Leonard	120	*Out of print*
37	James Cason	109	*Out of print*
38	John C. Sheehan	120	*Out of print*
39	Max Tishler	114	*Out of print*

Collective Vol. IV A revised edition of Annual Volumes 30–39
Norman Rabjohn, *Editor-in-Chief* 1036

40	Melvin S. Newman	114	*Out of print*
41	John D. Roberts	118	*Out of print*
42	Virgil Boekelheide	118	*Out of print*
43	B. C. McKusick	124	*Out of print*
44	The Late William E. Parham	131	*Out of print*
45	William G. Dauben	118	*Out of print*
46	E. J. Corey	146	*Out of print*
47	William D. Emmons	140	*Out of print*
48	Peter Yates	164	*Out of print*
49	Kenneth B. Wiberg	124	*Out of print*

Collective Vol. V A revised edition of Annual Volumes 40–49
Henry E. Baumgarten, *Editor-in-Chief* 1234

Cumulative Indices to Collective Volumes I, II, III, IV, V
Ralph L. and The Late Rachel H. Shriner, *Editors*

50	Ronald Breslow	136	*Out of print*
51	Richard E. Benson	209	*Out of print*
52	Herbert O. House	192	*Out of print*
53	Arnold Brossi	193	*Out of print*
54	Robert E. Ireland	155	*Out of print*
55	Satoru Masamune	150	*Out of print*
56	George H. Buchi	144	*Out of print*
57	Carl R. Johnson	135	*Out of print*
58	The Late William A. Sheppard	216	*Out of print*
59	Robert M. Coates	267	*Out of print*

Collective Vol. VI A revised edition of Annual Volumes 50–59
Wayland E. Noland, *Editor-in-Chief* 1208

60	Orville L. Chapman		
61	The Late Robert V. Stevens		
62	Martin F. Semmelhack		
63	Gabriel Saucy		
64	Andrew S. Kende		
65	Edwin Vedejs		
66	Clayton H. Heathcock		

Collective Volumes, Cumulative Indices, and Annual Volumes 60–65 are available from John Wiley & Sons, Inc.

ORGANIC SYNTHESES

ORGANIC SYNTHESES

AN ANNUAL PUBLICATION OF SATISFACTORY METHODS FOR THE PREPARATION OF ORGANIC CHEMICALS

VOLUME 66

1988

JOHN WILEY & SONS

NEW YORK • CHICHESTER • BRISBANE • TORONTO • SINGAPORE

Published by John Wiley & Sons, Inc.

"John Wiley & Sons, Inc. is pleased to publish this volume of Organic Syntheses on behalf of Organic Syntheses, Inc. Although Organic Syntheses, Inc. has assured us that each preparation contained in this volume has been checked in an independent laboratory and that any hazards that were uncovered are clearly set forth in the write-up of each preparation, John Wiley & Sons, Inc. does not warrant the preparations against any safety hazards and assumes no liability with respect to the use of the preparations."

Library of Congress Catalog Card Number: 21-17747
ISBN 0-471-61972-8

Printed in the United States of America

10 9 8 7 6 5 4 3 2 1

NOTICE

With Volume 62, the Editors of *Organic Syntheses* began a new presentation and distribution policy to shorten the time between submission and appearance of an accepted procedure. The soft-cover edition of this volume is produced by a rapid and inexpensive process, and is sent at no charge to members of the Organic Divisions of the American and French Chemical Societies, The Perkin Division of the Royal Society of Chemistry, and The Society of Synthetic Organic Chemistry, Japan. The soft-cover edition is intended as the personal copy of the owner and is not for library use. A hardcover edition is published by John Wiley and Sons, Inc. in the traditional format, and differs in content primarily in the inclusion of an index. The hardcover edition is intended primarily for library collections and is available for purchase through the publisher. Annual volumes 60–64 will be included in a new five-year version of the collective volumes of *Organic Syntheses* which will appear as *Collective Volume Seven* in the traditional hardcover format, after the appearance of annual volume 64. It will be available for purchase from the publishers. The Editors hope that the new *Collective Volume* series, appearing twice as frequently as the previous decennial volumes, will provide a permanent and timely edition of the procedures for personal and institutional libraries. The Editors welcome comments and suggestions from users concerning the new editions.

NOMENCLATURE

Both common and systematic names of compounds are used throughout this volume, depending on which the Editor-in-Chief felt was more appropriate. The *Chemical Abstracts* indexing name for each title compound, if it differs from the title name, is given as a subtitle. Systematic *Chemical Abstracts* nomenclature, used in both the 9th and 10th Collective Indexes for the title compound and a selection of other compounds mentioned in the procedure, is provided in an appendix at the end of each preparation. Registry numbers, which are useful in computer searching and identification, are also provided in these appendixes. Whenever two names are concurrently in use and one name is the correct *Chemical Abstracts* name, that name is preferred.

SUBMISSION OF PREPARATIONS

Organic Syntheses welcomes and encourages submission of experimental procedures which lead to compounds of wide interest or which illustrate important new developments in methodology. The Editorial Board will consider proposals in outline format as shown below, and will request full experimental details for those proposals which are of sufficient interest. Submissions which are longer than three steps from commercial sources or from existing *Organic Syntheses* procedures will be accepted only in unusual circumstances.

Organic Syntheses Proposal Format

1. Authors
2. Literature reference (enclose preprint if available)
3. Proposed sequence
4. Best current alternative(s)
5. a. Proposed scale, final product
   b. Overall yield
   c. Method of isolation and purification
   d. Purity of product (%)
   e. How determined
6. Any unusual apparatus or experimental technique

7. Any hazards
8. Source of starting material
9. Utility of method or usefulness of product

Submit to: Dr. Jeremiah P. Freeman, Secretary
Department of Chemistry
University of Notre Dame
Notre Dame, IN 46556

Proposals will be evaluated in outline form and again after submission of full experimental details and discussion. A procedure that has been accepted by The Editorial Board will not be published until the procedure has been satisfactorally reproduced and accepted for publication by an Editor. A form that details the preparation of a complete procedure (Notice to Submitters) can be obtained from the Secretary.

Additions, corrections, and improvements to the preparations previously published are welcomed; these should be directed to the Secretary. However, checking of such improvements will only be undertaken when new methodology is involved. Substantially improved procedures have been included in the Collective Volumes in place of a previously published procedure.

PREFACE

This volume contains 28 preparations that illustrate many of the most active areas of synthetic organic chemistry. It starts with a procedure for the preparation of **1-METHYL-1-(TRIMETHYLSILYL)ALLENE** and the use of this interesting reagent for **[3+2] ANNULATION.** These procedures are followed by another organosilicon procedure, for the preparation of the useful reagent **(1-OXO-2-PROPENYL)-TRIMETHYLSILANE.**

Claisen rearrangement of a propargyl alcohol is illustrated by the synthesis of the allenic ester **ETHYL 3,4-DECADIENOATE,** which is isomerized by treatment with alumina to form the $\alpha,\beta,\gamma,\delta$-unsaturated ester **ETHYL (E,Z)-2,4-DECADIENOATE.** Büchi's useful method for the formation of γ,δ-unsaturated aldehydes, Claisen rearrangement of (E)-allyloxyacrylic acids, is illustrated in the preparation of **3-PHENYL-4-PENTENAL.**

The aprotic Michael addition reaction is demonstrated in an imaginative fashion in the reaction of the dienolate of 3-methyl-2-cyclohexen-1-one with methyl crotonate to give the methyl ester of **1,3-DIMETHYL-5-OXOBICYCLO-[2.2.2]OCTANE-2-CARBOXYLIC ACID.** From enolates we turn to homoenolates in the **COPPER-CATALYZED CONJUGATE ADDITION OF A ZINC HOMOENOLATE.** The latter procedure, in which silicon, zinc, and copper all play major roles, vividly demonstrates the importance of metals in modern organic synthesis.

Further illustration of the utility of metals is provided in the next six procedures, involving copper, palladium, zinc, tin, silicon, and iron. This section of the volume begins with use of Pd(II) in the catalytic dimerization of methyl acrylate to **DIMETHYL (E)-2-HEXENEDIOATE,** which reacts with a "higher-order" cuprate (Lipshutz reagent) to provide **2-CARBOMETHOXY-3-VINYLCYCLOPENTANONE.** Following are procedures for the **PALLADIUM-CATALYZED SYNTHESIS OF CONJUGATED DIENES** and the **SYNTHESIS OF BIARYLS VIA PALLADIUM-CATALYZED CROSS COUPLING** of an arylzinc reagent with an aryl bromide. The former procedure is illustrated by the preparation of **(5Z,7E)-5,7-HEXADECADIENE** and the latter by the formation of **2-METHYL-4′-NITROBIPHENYL.** The next preparation, **VINYL RADICAL CYCLIZATION VIA ADDITION OF TIN RADICALS TO TRIPLE BONDS,** provides an example of the interesting radical cyclization chemistry that has recently been developed in Gilbert Stork's laboratory. Another organosilicon procedure follows, **CYCLOPENTANONES FROM CARBOXYLIC ACIDS VIA INTRAMOLECULAR ACYLATION OF ALKYLSILANES: 2-METHYL-2-VINYLCYCLOPENTANONE.** In addition to demonstrating the "alkyl Friedel–Crafts acylation," the latter procedure also provides an example of the α-alkylation of the lithium dienolate resulting from deprotonation of an α,β-unsaturated ester. The (cyclopentadienyl)(dicarbonyl)iron complex of ethyl vinyl ether is used as a vinyl cation synthon in the preparation of **trans-3-METHYL-2-VINYLCYCLOHEXANONE.** This interesting procedure also illustrates the electrophilic capture of the enolate resulting from the conjugate addition of a cuprate reagent to an enone.

A novel annulation process is demonstrated in Ley's **PREPARATION OF tert-BUTYL ACETOTHIOACETATE** and its use in the synthesis of **3-ACETYL-4-HYDROXY-5,5-DIMETHYLFURAN-2(5H)-ONE.** Following are procedures for the preparation of ketones and aldehydes from carboxylic acids, which are demonstrated with the copper-catalyzed reaction of a Grignard reagent with an acyl chloride in the preparation of **METHYL 6-OXODECANOATE** and the reduction of an acyl chloride with lithium (tri-tert-butoxy)aluminum hydride, giving **6-OXODECANAL.**

The acetylene "zipper reaction" is illustrated by the conversion of **2-DECYN-1-OL** to **9-DECYN-1-OL** by the use of a mixture of potassium tert-butoxide and lithiated 1,3-diaminopropane. The next procedure involves the unusual reagent **[I,I-BIS(TRIFLUOROACETOXY)]-IODOBENZENE** in the formation of **CYCLOBUTYLAMINE HYDROCHLORIDE FROM CYCLOBUTANECARBOXAMIDE,** a version of the Hofmann rearrangement that proceeds under mildly acidic conditions. The **INVERSE ELECTRON DEMAND DIELS-ALDER REACTION OF AN ELECTRON-DEFICIENT HETEROCYCLIC AZADIENE** is illustrated by the preparation of **TRIETHYL 1,2,4-TRIAZINE-3,5,6-TRICARBOXYLATE.**

Stereospecific diazotization is used for the preparation of **(S)-2-CHLOROALKANOIC ACIDS OF HIGH ENANTIOMERIC PURITY FROM (S)-2-AMINO ACIDS.** The (S)-2-chloropropanoic acid produced in this process is reduced by lithium aluminum hydride to give (S)-2-chloropropan-1-ol, which is cyclized by base to give the optically active epoxide in **(R)-ALKYLOXIRANES OF HIGH ENANTIOMERIC PURITY FROM (S)-2-CHLOROALKANOIC ACIDS VIA (S)-2-CHLORO-1-ALKANOLS: (R)-METHYLOXIRANE.**

A version of the Grob fragmentation reaction is used to prepare an α,β-acetylenic ester in **UTILIZATION OF β-CHLORO ALKYLIDENE/ARYLIDENE MALONATES IN THE SYNTHESIS OF ETHYL CYCLOPROPYLPROPIOLATE.** Another uncommon process is illustrated in **OXIDATIVE CLEAVAGE OF AN AROMATIC RING: cis,cis-MONOMETHYL MUCONATE FROM 1,2-DIHYDROXYBENZENE.** A novel twist on the Beckmann rearrangement is demonstrated in **PREPARATION OF 2-PROPYL-1-AZACYCLOHEPTANE FROM CYCLOHEXANONE OXIME.** The next procedure is a preparation of **6-DIETHYLPHOSPHONOMETHYL-2,2-DIMETHYL-1,3-DIOXEN-4-ONE,** a useful reagent for the synthesis of acyl tetronic and tetramic acids. Following is a recipe for the preparation of the Davis oxidizing agent, **(±)-trans-2-(PHENYLSULFONYL)-3-PHENYLOXAZIRIDINE.** The procedure for the preparation of the important reagent **TRISAMMONIUM GERANYL DIPHOSPHATE** utilizes techniques not often used in organic chemistry laboratories, lyophyilization and ion exchange chromatography. The volume concludes with a preparation of **ETHYL α-(HYDROXYMETHYL)ACRYLATE,** a useful intermediate for the preparation of bis-electrophiles such as ethyl α-(bromomethyl)acrylate.

The Editors of *Organic Syntheses* welcome communications from the chemical community about important preparations or procedures that might be included in future volumes. These communications may take the form of an actual proposed submission (see the Organic Syntheses Proposal Format at the end of the paperback version of this volume) or suggestions about important reactions that the Board might solicit from some third party. Keep in mind that the philosophy of the publication is dual—to provide reliable procedures for the preparation of specific compounds and to demonstrate general procedures. If the purpose is to demonstrate a procedure, it is nevertheless important to select an application for which the demonstrated procedure is superior to other available methods for the specific product.

The Editors of *Organic Syntheses* continue to benefit from the outstanding service of Professor Jeremiah P. Freeman, our Secretary, and Dr. Theodora W. Greene, our Assistant Editor.

The structures were prepared with the ChemDraw program.

CLAYTON H. HEATHCOCK

Berkeley, California
June, 1987

HENRY GILMAN
May 9, 1893–November 7, 1986

Henry Gilman was one of the dominant figures in American organic chemistry of the 20th century. A man of exceptional skill and industry, he made prodigious and seminal contributions to [illegible] at the time of his death he was 93 and the oldest living member of the Advisory Board of Organic Syntheses. He joined the Editorial Board in 1928 and was Editor-in-Chief in 1936.

Gilman was born in Boston on May 9, 1893 and graduated summa cum laude from Harvard in 1915, working for part of his undergraduate period in close association with Roger Adams, who was then an instructor. Gilman obtained his A.M. and Ph.D. at Harvard with E. P. Kohler in

HENRY GILMAN
May 9, 1893–November 7, 1986

Henry Gilman was one of the dominant figures in American organic chemistry of the 20th century. A man of exceptional will and foresight who made prodigious and seminal contributions to chemistry, at the time of his death he was 93 and the oldest living member of the Advisory Editorial Board of *Organic Syntheses*. He joined the Editorial Board in 1924 and was Editor-in-Chief in 1926.

Gilman was born in Boston on May 9, 1893 and graduated *summa cum laude* from Harvard in 1915, working for part of his undergraduate period in close association with Roger Adams, who was then instructor. Gilman obtained his A.M. and Ph.D. at Harvard with E. P. Kohler in

1917 and 1918, respectively, and also spent some time at Zurich with Staudinger, as well as brief interludes at the Sorbonne and Oxford. Gilman was instructor at Harvard in 1917–1918 and then Associate at Illinois before moving to Iowa State in 1919, where he spent the rest of his academic career.

Gilman's greatest contributions were to organometallic chemistry and he worked in this field in the broadest sense. Starting with Grignard reagents, he covered the periodic table rather generally from lithium to uranium, back in the days when there were few, if any, glove boxes and almost no good way to characterize highly reactive substances, except by their reaction products. Many organic chemists have used the Gilman color test for formation of Grignard reagents and employed his procedures for the preparation and reactions of organolithium compounds.

Less well-known is his early work on cadmium and copper compounds; the latter, in the form of cuprates, have been adapted in many laboratories for use in synthetic procedures for many otherwise difficultly accessible substances. Although much early work was done by F. S. Kipping and F. C. Whitmore on silicon compounds, Gilman made very substantial contributions to this field as well, and these were recognized by the first Kipping award. Gilman also made the initial discoveries of rearrangements in nucleophilic aromatic substitution reactions by lithium amides, which were later demonstrated to involve arynes as intermediates. Another of his important interests was in heterocyclic chemistry, especially the chemistry of furans and thiophenes. All in all, he published just over 1000 research papers. The multivolume treatise, *Organic Chemistry,* which ran through several editions, starting in 1938, was the bible for several generations of graduate students studying for their written or oral examinations.

Gilman epitomized the work ethic in organic research. Not only did he work hard himself, he expected at least as much from his students. With quite a reputation as a laboratory slave driver, he turned out a coterie of very well-trained and highly successful students, while at the same time gaining their respect and affection.

Besides the Kipping Award, Gilman received many honors; among them were membership in the National Academy of Sciences, foreign member of the Royal Society, the Midwest Award and the Priestly Medal of the American Chemical Society.

That half of his papers were published after he lost almost all of his sight as the result of a detached retina and glaucoma in 1947 is quite a tribute to his will and tenacity, both of which were greatly reinforced by the substantial efforts of Ruth, his charming wife of 57 years, who literally acted as his eyes for almost forty years of their life together. Henry Gilman died on November 7, 1986 and his wife just shortly thereafter on January 28, 1987. His many contributions to Iowa State University are fittingly memorialized by a chemistry building, Gilman Hall, by a Gilman Graduate Fellowship Fund, and by annual Gilman Lectures.

JOHN D. ROBERTS

September 11, 1987

CONTENTS

A GENERAL METHOD FOR THE SYNTHESIS OF ALLENYLSILANES: 1-METHYL-1-(TRIMETHYLSILYL)ALLENE

(Silane, trimethyl(1-methyl-1,2-propadienyl)-)

$$HOCH_2C{\equiv}CSiMe_3 \xrightarrow[2.\ CH_3SO_2Cl]{1.\ CH_3MgCl,\ THF} CH_3SO_3CH_2C{\equiv}CSiMe_3$$

$$CH_3SO_3CH_2C{\equiv}CSiMe_3 \xrightarrow[LiBr,\ THF]{CH_3MgCl,\ CuBr,} H_2C{=}C{=}C(SiMe_3)(CH_3)$$

Submitted by Rick L. Danheiser, Yeun-Min Tsai, and David M. Fink.[1]

Checked by Marianne Marsi and Bruce E. Smart.

1. Procedure

A 500-mL, three-necked, round-bottomed flask is equipped with a magnetic stirring bar, rubber septum, low temperature thermometer, and a 250-mL pressure-equalizing dropping funnel fitted with a nitrogen inlet adapter (Note 1). The flask is charged with 30.0 g (0.234 mol) of 3-trimethylsilyl-2-propyn-1-ol (Note 2) and 230 mL of dry tetrahydrofuran (Note 3), and then cooled with an ice bath while 84 mL of a 2.8 M solution of methylmagnesium chloride in tetrahydrofuran (Note 4) is added at such a rate that the internal temperature does not rise above 10°C. Approximately 1.5 hr is required for the addition, after which time the gray solution is stirred at 0°C for 30 min, and then cooled below -70°C with a dry ice-acetone bath. Methanesulfonyl chloride (26.8 g, 0.234 mol) (Note 5) is added over 10 min via syringe, and after 30 min the cold bath is removed and the pale yellow reaction mixture is allowed to warm to room temperature over 2 hr.

A 2-L, three-necked, round-bottomed flask equipped with a vacuum adapter and two glass stoppers (Note 1) is charged with 21.4 g (0.246 mol) of anhydrous lithium bromide and 35.3 g (0.246 mol) of anhydrous cuprous bromide (Note 6). The reaction vessel is evacuated and the contents are briefly heated with a Bunsen burner flame. After 30 min the vacuum is replaced by nitrogen and the apparatus is equipped with a mechanical stirrer and two rubber septa. Dry tetrahydrofuran (260 mL) (Note 3) is added, and the resulting green solution containing a small amount of undissolved solid is cooled with an ice bath while 84 mL of a 2.8 M solution of methylmagnesium chloride in tetrahydrofuran (Note 4) is added rapidly via syringe over 1-2 min. After 20 min of further stirring at 0°C, the reaction mixture appears as a viscous yellow-green suspension. The solution of the mesylate derivative of 3-trimethylsilyl-2-propyn-1-ol prepared above is now transferred via cannula over 45 min to the reaction mixture, which is cooled below -70°C with a dry ice-acetone bath. After 30 min, the cold bath is removed and the green reaction mixture is stirred at room temperature for 2 hr. The blue-gray mixture is then poured into a 2-L Erlenmeyer flask containing a magnetically stirred mixture of 400 mL of pentane, 200 mL of water, and 400 mL of saturated ammonium chloride solution. The organic phase is separated and washed successively with two 200-mL portions of saturated ammonium chloride solution, ten 1-L portions of water (Note 7), and 100 mL of saturated sodium chloride solution. The organic phase is dried over anhydrous sodium sulfate, and the drying agent is removed by filtration. The solvent is removed from the filtrate by atmospheric distillation through a 10-cm Vigreux column. The residual liquid is carefully distilled through a 12-cm column packed with glass helices to give 21.3-22.2 g, (72-75%) of 1-methyl-1-(trimethylsilyl)-allene as a colorless liquid, bp 111°C (Notes 8-11).

2. Notes

1. The apparatus is flame-dried at 20 mm pressure and then maintained under an atmosphere of nitrogen during the course of the reaction.

2. 3-Trimethylsilyl-2-propyn-1-ol was obtained from Petrarch Systems, Inc. and used as received. Alternatively, it can be prepared by the silylation of 2-propyn-1-ol.[2]

3. Tetrahydrofuran was distilled from sodium benzophenone ketyl immediately before use.

4. Methylmagnesium chloride in tetrahydrofuran was purchased from Aldrich Chemical Company, Inc.

5. Methanesulfonyl chloride was obtained from the Aldrich Chemical Company, Inc. and purified by distillation from phosphorus pentoxide before use.

6. Lithium bromide, obtained from Aldrich Chemical Company, Inc., and cuprous bromide, supplied by Fluka Chemical Corporation, were dried at 120°C (0.02 mm) for 8 hr before use. The checkers obtained lower yields (54-58%) with cuprous bromide that was supplied by other commercial sources.

7. This procedure conveniently removes tetrahydrofuran from the organic phase.

8. The submitters report bp 112-113°C and state that an additional 2.2 g (7%) of product, bp 54-56°C (90 mm), can be obtained by combining the distillation forerun with the pot residue and redistilling the mixture at reduced pressure.

9. The allenylsilane thus obtained was found by ^{1}H NMR analysis to be contaminated with 7-8% of 1-trimethylsilyl-1-butyne. This material is suitable for use in [3+2] annulations.[3] Efficient stirring during the formation of the mixed cuprate reagent is important in minimizing the amount of this impurity produced in the reaction.

10. The submitters note that if pure 1-methyl-1-(trimethylsilyl)allene is required, the mixture of allenylsilane and 1-trimethylsilyl-1-butyne is treated with 0.15 equiv of silver nitrate in 10:1 methanol-water at room temperature for 1 hr. Extraction with pentane and distillation furnishes the allenylsilane in 79% yield. Gas chromatographic analysis (0.2 mm x 10.5 m methyl silicone-coated fused silica capillary column, split ratio 100:1, column pressure 20 psi, temperature 20°C) indicates that this material contains none of the acetylenic side product.

11. The product exhibits the following spectral properties: IR (neat) cm^{-1}: 2955, 2920, 2900, 2860, 1935, 1440, 1400, 1250, 935, 880, 830, 805, 750, and 685; ^{1}H NMR (250 MHz, $CDCl_3$) δ: 0.08 (s, 9 H), 1.67 (t, 3 H, J = 3.3), and 4.25 (q, 2 H, J = 3.3); ^{13}C NMR (67.9 MHz, $CDCl_3$) δ: -2.1, 15.1, 67.3, 89.1, and 209.1. The impurity, 1-trimethylsilyl-1-butyne, displays ^{1}H NMR peaks at 0.14 (s, 9 H), 1.15 (t, 3 H, J = 7), and 2.23 (q, 2 H, J = 7).

3. Discussion

Allenylsilanes serve as valuable three-carbon components in a general [3+2] annulation method for the synthesis of five-membered rings.[4] A variety of general synthetic approaches to allenylsilanes have recently been developed,[5-7] and a number of specialized routes to various specific functionalized derivatives[8] are available as well. The present procedure involves a modification of the general method of Vermeer,[5a] which is the most widely applicable route to substituted allenylsilanes. Advantages of this approach include its efficiency, high regioselectivity, and utility in the synthesis of a variety of mono-, di-, and trisubstituted allenylsilane derivatives.

1. Department of Chemistry, Massachusetts Institute of Technology, Cambridge, MA 02139.
2. Komarov, N. V.; Shostakovskii, M. F. *Izvest. Akad. Nauk, S.S.S.R. Otdel. Khim. Nauk* **1960**, 1300; *Chem. Abstr.* **1961**, *55*, 358g.
3. See Danheiser, R. L.; Fink, D. M.; Tasi, Y.-M. *Org. Synth.* **1988**, *66*, 8.
4. (a) Danheiser, R. L.; Carini, D. J.; Basak, A. *J. Am. Chem. Soc.* **1981**, *103*, 1604; (b) Danheiser, R. L.; Carini, D. J.; Fink, D. M.; Basak, A. *Tetrahedron* **1983**, *39*, 935.
5. (a) Westmijze, H.; Vermeer, P. *Synthesis* **1979**, 390; see also (b) Tanigawa, Y.; Murahashi, S.-I. *J. Org. Chem.* **1980**, *45*, 4536; (c) Montury, M.; Psaume, B.; Goré, J. *Tetrahedron Lett.* **1980**, *21*, 163.
6. Danheiser, R. L.; Carini, D. J. *J. Org. Chem.* **1980**, *45*, 3925.
7. (a) Patrick, T. B.; Haynie, E. C.; Probst, W. J. *J. Org. Chem.* **1972**, *37*, 1553; (b) Beard, C. D.; Craig, J. C. *J. Am. Chem. Soc.* **1974**, *96*, 7950; (c) Creary, X. *J. Am. Chem. Soc.* **1977**, *99*, 7632; (d) Yogo, T.; Koshino, J.; Suzuki, A. *Tetrahedron Lett.* **1979**, *20*, 1781; (e) Montury, M.; Gore, J. *Tetrahedron Lett.* **1980**, *21*, 51; (f) Fleming, I.; Newton, T. W.; Roessler, F. *J. Chem. Soc., Perkin Trans. I* **1981**, 2527; (g) Clinet, J.-C.; Linstrumelle, G. *Synthesis* **1981**, 875.

8. (a) Ruden, R. A. *J. Org. Chem.* **1974**, *39*, 3607; (b) Ganem, B. *Tetrahedron Lett.* **1974**, 4467; (c) Orlov, V. Yu.; Lebedev, S. A.; Ponomarev, S. V.; Lutsenko, I. F. *Zh. Obshch. Khim.* **1975**, *45*, 708; *Chem. Abstr.* **1975**, *83* 43428b; (d) Shen, C. C.; Ainsworth, C. *Tetrahedron Lett.* **1979**, 83, 87, 89, 93; (e) Bertrand, M.; Dulcere, J.-P.; Gil, G. *Tetrahedron Lett.* **1980**, *21*, 1945; (f) Clinet, J.-C.; Linstrumelle, G. *Tetrahedron Lett.* **1980**, *21*, 3987; (g) Mukaiyama, T.; Harada, T. *Chem. Lett.* **1981**, 621; (h) Daniels, R. G.; Paquette, L. A. *Tetrahedron Lett.* **1981**, *22*, 1579; (i) Visser, R. G.; Brandsma, L.; Bos, H. J. T. *Tetrahedron Lett.* **1981**, *22*, 2827; (j) Dulcere, J.-P.; Grimaldi, J.; Santelli, M. *Tetrahedron Lett.* **1981**, *22*, 3179; (k) Pillot, J.-P.; Bennetau, B.; Dunogues, J.; Calas, R. *Tetrahedron Lett.* **1981**, *22*, 3401; (l) Fleming, I.; Perry, D. A. *Tetrahedron* **1981**, *37*, 4027; (m) Kruithof, K. J. H.; Klumpp, G. W. *Tetrahedron Lett.* **1982**, *23*, 3101; (n) Pornet, J.; Mesnard, D.; Miginiac, L. *Tetrahedron Lett.* **1982**, *23*, 4083; (o) Parker, K. A.; Petraitis, J. J.; Kosley, R. W.; Buchwald, S. L. *J. Org. Chem.* **1982**, *47*, 389; (p) Ishiguro, M.; Ikeda, N.; Yamamoto, H. *J. Org. Chem.* **1982**, *47*, 2225; (q) Verkruijsse, H. D.; Verboom, W.; Van Rijn, P. E.; Brandsma, L. *J. Organomet. Chem.* **1982**, *232*, C1; (r) Van Rijn, P. E.; Brandsma, L. *J. Organomet. Chem.* **1982**, *233*, C25; (s) Barton, T. J.; Hussmann, G. P. *J. Am. Chem. Soc.* **1983**, *105*, 6316; (t) Wang, K. K.; Nikam, S. S.; Ho, C. D. *J. Org. Chem.* **1983**, *48*, 5376; (u) Cutting, I.; Parsons, P. J. *Tetrahedron Lett.* **1983**, *24*, 4463; (v) Bridges, A. J.; Fedij, V.; Turowski, E. C. *J. Chem. Soc., Chem. Commun.* **1983**, 1093; (w) Reich, H. J.; Kelly, M. J.; Olson, R. E.; Holtan, R. C. *Tetrahedron* **1983**, *39*, 949.

Appendix

Chemical Abstracts Nomenclature (Collective Index Number); (Registry Number)

1-Methyl-1-(trimethylsilyl)allene: Silane, trimethyl(1-methyl-1,2-propadienyl)- (10); (74542-82-8)

3-Trimethylsilyl-2-propyn-1-ol: 2-Propyn-1-ol, 3-(trimethylsilyl)- (9); (5272-36-6)

A GENERAL [3+2] ANNULATION: CIS-4-EXO-ISOPROPENYL-1,9-DIMETHYL-8-(TRIMETHYLSILYL)BICYCLO[4.3.0]NON-8-EN-2-ONE

(4H-Inden-4-one, 1,3a,5,6,7,7a-hexahydro-3,3a-dimethyl-6-(1-methylethenyl)-2-(trimethylsilyl)-, (3aα,6α,7aα)-)

Submitted by Rick L. Danheiser, David M. Fink, and Yeun-Min Tsai.[1]

Checked by Marianne Marsi and Bruce E. Smart.

1. Procedure

A 500-mL, three-necked, round-bottomed flask is equipped with a 25-mL pressure-equalizing dropping funnel, a mechanical stirrer, and a Claisen adapter fitted with a nitrogen inlet adapter and a low temperature thermometer (Note 1). The flask is charged with 11.5 g (0.077 mol) of (R)-(-)-carvone (Note 2), 10.8 g (0.079 mol) of 1-methyl-1-(trimethylsilyl)allene (Note 3), and 180 mL of dry dichloromethane (Note 4), and then cooled below -75°C with a dry ice-acetone bath while a solution of 17.4 g (0.092 mol) of titanium tetrachloride (Note 5) in 10 mL of dichloromethane is added dropwise over 1 hr. After 30 min, the cold bath is removed, and the reaction mixture, which appears as a red suspension, is allowed to warm to 0°C over approximately 30 min. The resulting dark red solution is poured slowly into a 2-L Erlenmeyer flask containing a magnetically-stirred mixture of 400 mL of diethyl ether and 400 mL of water (Note 6). The aqueous phase is separated and extracted with two 200-mL portions of ether. The combined organic phases are washed with 250 mL of water and 250 mL of saturated sodium chloride solution, dried over anhydrous magnesium sulfate, and concentrated at reduced pressure using a rotary evaporator. The residual yellow liquid is distilled through a 15-cm Vigreux column at reduced pressure to afford 17.5 g (82%) of the bicyclononenone **1** as a very pale yellow liquid, bp 98-101°C (0.03 mm), $[\alpha]_D^{20}$ -157.8 ± 0.8 (1.57, CH_2Cl_2) (Notes 7 and 8).

2. Notes

1. The apparatus is flame-dried under vacuum and then maintained under an atmosphere of nitrogen during the course of the reaction.

2. (R)-(-)-Carvone was purchased from Aldrich Chemical Company, Inc. and distilled before use.

3. 1-Methyl-1-(trimethylsilyl)allene (90% purity, contaminated with 10% 1-trimethylsilyl-1-butyne) was prepared by the method of Danheiser, R. L.; Tsai, Y. M.; Fink, D. M. *Org. Synth.* **1988**, *66*, 1.

4. Dichloromethane was distilled from calcium hydride immediately before use.

5. Titanium tetrachloride (99.9%) was obtained from the Aldrich Chemical Company, Inc. and distilled before use. Lower yields (70-77%) resulted if the titanium tetrachloride was not distilled. Unreacted carvone is recovered if a small excess of titanium tetrachloride is not used.

6. The two-phase mixture is vigorously stirred using a 7-cm Teflon-coated magnetic stirring bar.

7. The submitters report obtaining 18.8 g (88%) of product, bp 93-96°C (0.03 mm). The purity of the product was determined to be >99% by gas chromatographic analysis (10% OV-101 on 100-120 mesh Chromosorb W, 6 ft x 1/8 in, program: 200°C for 2 min and then 200-300°C at 32°C/min).

8. The product exhibits the following spectral properties: IR (neat) cm^{-1}: 3080, 2950, 2920, 1700, 1640, 1610, 1440, 1375, 1315, 1245 1200, 830, 755, 680; ^1H NMR (250 MHz, $CDCl_3$) δ: 0.08 (s, 9 H), 1.10 (s, 3 H), 1.65 (t, 3 H, J = 2.2), 1.68 (m, 3 H), 1.65-1.72 (m, 2 H), 2.10 (d of m, 1 H, J = 12.4), 2.15-2.25 (m, 1 H), 2.27-2.31 (m, 2 H), 2.45-2.57 (m, 2 H), 4.63 (m, 1 H), 4.72 (m, 1 H); ^{13}C NMR (62.8 MHz, $CDCl_3$) δ: -0.7, 14.4, 21.1, 21.3, 32.4, 39.8, 42.0, 43.9, 46.1, 64.8, 110.4, 136.1, 147.7, 151.5, 215.6.

3. Discussion

The procedure described here serves to illustrate a general [3+2] annulation method[2] for the synthesis of cyclopentane derivatives. A unique feature of this one-step annulation is its capacity to generate regio-specifically five-membered rings substituted at each position, and functionally equipped for further synthetic elaboration. As formulated in the following equation, the reaction proceeds with remarkably high stereo-selectivity via the effective suprafacial addition of the three-carbon allene component to an electron-deficient olefin ("allenophile").

R W RE RZ + R2 R3 C=C=C SiMe3 R1 —TiCl4→ R R1 W SiMe3 RZ RE R3 R2

Some representative examples of the [3+2] annulation are listed in Table I. Both cyclic and acyclic allenophiles participate in the reaction. α-Alkylidene ketones undergo annulation to provide access to spiro-fused systems, and acetylenic allenophiles react to form cyclopentadiene derivatives. The reactions of (E)- and (Z)-3-methyl-3-penten-2-one illustrate the stereochemical course of the annulation, which proceeds with a strong preference for the suprafacial addition of the allene to the two-carbon allenophile. The high stereoselectivity displayed by the reaction permits the stereocontrolled synthesis of a variety of mono- and polycyclic systems.

1. Department of Chemistry, Massachusetts Institute of Technology, Cambridge, MA 02139.

2. (a) Danheiser, R. L.; Carini, D. J.; Basak, A. *J. Am. Chem. Soc.* **1981**, *103*, 1604; (b) Danheiser, R. L.; Carini, D. J.; Fink, D. M.; Basak, A. *Tetrahedron* **1983**, *39*, 935.

TABLE I

[3+2] ANNULATIONS EMPLOYING ALLENYLSILANES

Allenophile	Allene	Annulation Product	% Yield
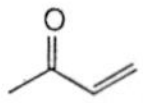	$H_2C{=}C{=}C(SiMe_3)Me$	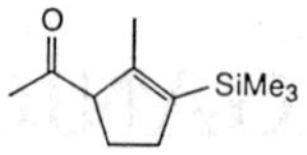 SiMe3	71-75
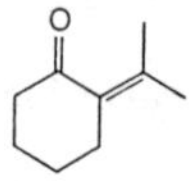		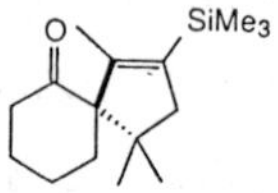 SiMe3	86
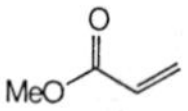		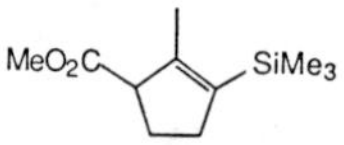 MeO2C, SiMe3	49
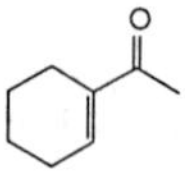		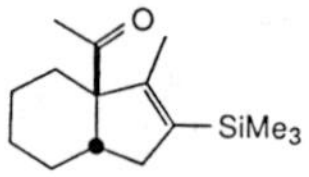 SiMe3	91
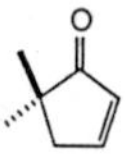		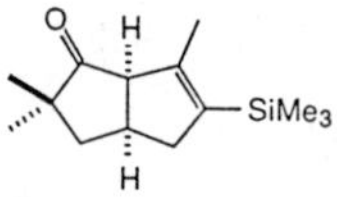 H, H, SiMe3	90
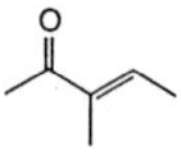		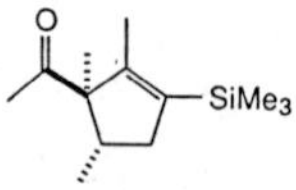 SiMe3	71
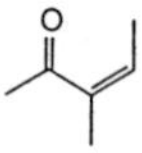		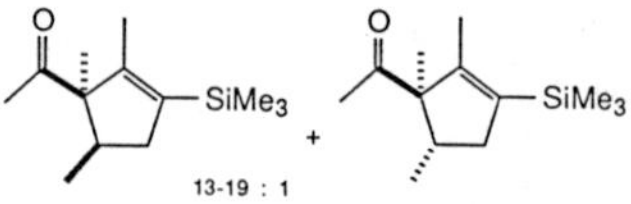 SiMe3 + SiMe3 (13-19 : 1)	68
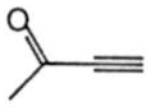	$Me_2C{=}C{=}C(SiMe_3)Me$	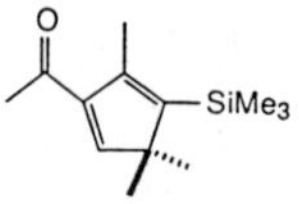 SiMe3	53
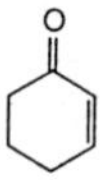		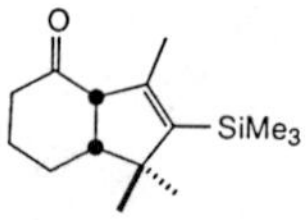 SiMe3	63

Appendix

Chemical Abstracts Nomenclature (Collective Index Number); (Registry Number)

cis-4-exo-Isopropenyl-1,9-dimethyl-8-(trimethylsilyl)bicyclo[4.3.0]non-8-en-2-one: 4H-Inden-4-one, 1,3a,5,6,7,7a-hexahydro-3,3a-dimethyl-6-(1-methylethenyl)-2-(trimethylsilyl)-, (3aα,6α,7aα)- (10); (77494-23-6)

(R)-(-)-Carvone: p-Mentha-6,8-dien-2-one, (R)-(-)- (8); 2-Cyclohexen-1-one, 2-methyl-5-(1-methylethenyl)-, (R)- (9); (6485-40-1)

1-Methyl-1-(trimethylsilyl)allene: Silane, trimethyl(1-methyl-1,2-propadienyl)- (10); (74542-82-8)

(1-OXO-2-PROPENYL)TRIMETHYLSILANE

(Silane, trimethyl(1-oxo-2-propenyl)-)

Submitted by Rick L. Danheiser, David M. Fink, Kazuo Okano, Yeun-Min Tsai, and Steven W. Szczepanski.[1]

Checked by Masahiko Hayashi and Ryoji Noyori.

1. Procedure

A. *(1-Hydroxy-2-propenyl)trimethylsilane*. A 2-L, three-necked, round-bottomed flask is equipped with a magnetic stirring bar, two pressure-equalizing dropping funnels (250 and 500 mL), and a Claisen adapter fitted with an argon inlet adapter and a rubber septum (Note 1). The flask is charged with 20.0 g (0.344 mol) of allyl alcohol (Note 2), and 400 mL of dry tetrahydrofuran (Note 3), and then cooled below -75°C with a dry-ice acetone bath and maintained at that temperature while 157 mL (0.363 mol) of a 2.31 M solution of n-butyllithium in hexane (Note 4) is added dropwise over 1 hr. After 50 min, a solution of 39.3 g (0.362 mol) of chlorotrimethylsilane (Note 5) in 25 mL of tetrahydrofuran is added dropwise via syringe over 30 min, and the resulting colorless reaction mixture is stirred for 1 hr further, and then treated dropwise over 1.5 hr with 258 mL (0.415 mol) of a 1.61 M solution of tert-butyllithium in pentane (Note 4). After 2 hr of further stirring at -75°C the cold bath is removed, and 100 mL of saturated ammonium chloride solution is added in one portion to the yellow reaction mixture. The resulting solution is stirred for 5 min, and then diluted with 50 mL of water and 300 mL of pentane. The organic phase is separated and washed successively with three 100-mL portions of water and two 100-mL portions of saturated sodium chloride solution, dried over anhydrous sodium sulfate, filtered, and concentrated by carefully distilling off the solvents at atmospheric pressure through a 10-cm Vigreux column. The residual pale yellow liquid is transferred to a 100-mL round-bottomed flask, and the remaining volatile impurities are removed by distillation at 15 mm through a 4-cm column packed with glass helices (Note 6), leaving 35.1-39.7 g of (1-hydroxy-2-propenyl)trimethylsilane as a pale yellow liquid (Notes 7 and 8) used in the next step without further purification.

B. (1-Oxo-2-propenyl)trimethylsilane. A 2-L, three-necked, round-bottomed flask is equipped with a mechanical stirrer and two 250-mL pressure-equalizing dropping funnels, one of which is fitted with an argon inlet adapter (Note 1). The flask is charged with 41.71 g (0.329 mol) of oxalyl chloride (Note 9) and 500 mL of dichloromethane (Note 10), and cooled below -75°C with a dry ice-acetone bath and maintained at that temperature while a solution of 55.82 g (0.715 mol) of dimethyl sulfoxide (Note 11) in 60 mL of dichloromethane is added dropwise over 1 hr. After 1 hr, a solution of the crude (1-hydroxy-2-propenyl)trimethylsilane in 100 mL of dichloromethane is added dropwise over 1.25 hr to the colorless reaction mixture, which is stirred at -75°C for 1 hr further, and then treated dropwise over 30 min with 150.38 g (1.486 mol) of triethylamine (Note 12). After 1 hr, the cold bath is removed and the reaction mixture is poured into 200 mL of water. The organic phase is separated and washed successively with five 100-mL portions of 10% hydrochloric acid, three 100-mL portions of water, and two 100-mL portions of saturated sodium chloride solution, dried over anhydrous sodium sulfate, filtered, and concentrated by carefully distilling off the solvents at atmospheric pressure through a 10-cm Vigreux column. The residual yellow oil is transferred to a 250-mL, round-bottomed flask containing 0.050 g of 3-tert-butyl-4-hydroxy-5-methylphenyl sulfide (Note 13) and distilled through a 4-cm column packed with glass helices to afford 27.8-30.0 g (63-68% overall yield based on allyl alcohol) of (1-oxo-2-propenyl)trimethylsilane as a brilliant yellow oil, bp 47-50°C (30 mm) (Notes 14 and 15).

2. Notes

1. The glass components of the apparatus are dried overnight in a 120°C oven, and then assembled and maintained under an atmosphere of argon during the course of the reaction.

2. Allyl alcohol was purchased from Aldrich Chemical Company, Inc. and distilled from calcium hydride prior to use.

3. Tetrahydrofuran was distilled from sodium benzophenone ketyl immediately before use.

4. n-Butyllithium was purchased from Aldrich Chemical Company, Inc. or Mitsuwa Pure Chemicals. tert-Butyllithium was obtained from Aldrich Chemical Company, Inc. These were titrated using the method of Watson and Eastham[2a] (submitters) or Lipton[2b] (checkers).

5. Chlorotrimethylsilane was obtained from Petrarch Systems, Inc., or Shin-etsu Kagaku Co., and distilled from calcium hydride before use.

6. The heating bath temperature was not permitted to exceed 70°C during the course of the distillation.

7. The purity of this material was determined to be 95% by gas chromatographic analysis (10% OV-101 on 100-120 mesh Chromosorb W, 6 ft x 1/8 in, program: 50°C for 2 min and then 50-250°C at 32°C/min).

8. The product exhibits the following spectral properties: IR (film) cm^{-1}: 3420, 2955, 2895, 2820, 1625, 1410, 1245, 1140, 1095, 990, 900, and 840; ^{13}C NMR (67.9 MHz, $CDCl_3$) δ: -4.4, 68.9, 109.4, 139.9; ^{1}H NMR (250 MHz, $CDCl_3$) δ: 0.05 (s, 9 H), 2.86 (br s, 1 H), 3.88 (m, 1 H), 4.86 (ddd, 1 H, J = 2, 2, 11), 4.98 (ddd, 1 H, J = 2, 2, 17), 5.89 (ddd, 1 H, J = 5.5, 11, 17); HRMS m/e calcd for $C_6H_{14}OSi$ (M^+): 130.0814, found: 130.0810.

9. Oxalyl chloride purchased from Aldrich Chemical Company, Inc. was fractionally distilled under argon before use.

10. Dichloromethane was distilled from calcium hydride immediately before use.

11. Dimethyl sulfoxide was distilled from calcium hydride immediately before use.

12. Triethylamine was distilled from calcium hydride before use.

13. 3-tert-Butyl-4-hydroxy-5-methylphenyl sulfide was purchased from Aldrich Chemical Company, Inc.

14. The purity of this material was determined to be >97% by gas chromatographic analysis (10% OV-101 on 100-120 mesh Chromosorb W, 6 ft x 1/8 in, program: 50°C for 2 min and then 50-250°C at 32°C/min).

15. The product exhibits the following spectral properties: IR (film) cm^{-1}: 2960, 2900, 1635, 1600, 1590, 1415, 1390, 1255, 1185, 985, 960, and 845; ^{13}C NMR (67.9 MHz, $CDCl_3$) δ: -2.2, 128.5, 141.3, 237.9; ^{1}H NMR (250 MHz, $CDCl_3$) δ: 0.23 (s, 9 H), 5.94 (dd, 1 H, J = 1, 11), 6.13 (dd, 1 H, J = 1, 18), 6.38 (dd, 1 H, J = 11, 18).

3. Discussion

(1-Oxo-2-propenyl)trimethylsilane has previously been prepared by Reich and co-workers in four steps beginning with propargyl alcohol.[3] This earlier synthesis proceeded in 45% overall yield and involved as key steps the metalation (at -90°C) and silylation of 1-(1-ethoxyethoxy)-1,2-propadiene, followed by careful hydrolysis of the resulting α-silyl allenyl ether.

The present method[4] offers a more efficient and convenient two-step route to the parent α,β-unsaturated acylsilane derivative. The first step in the procedure involves the conversion of allyl alcohol to allyl trimethylsilyl ether, followed by metalation[5] (in the same flask) with tert-butyllithium at -75°C. Protonation of the resulting mixture of interconverting lithium derivatives (**2** and 3) with aqueous ammonium chloride solution furnishes (1-hydroxy-2-propenyl)trimethylsilane (**4**), which is smoothly transformed to (1-oxo-2-propenyl)trimethylsilane by Swern oxidation.[6] The acylsilane is obtained in 63-68% overall yield from allyl alcohol in this fashion.

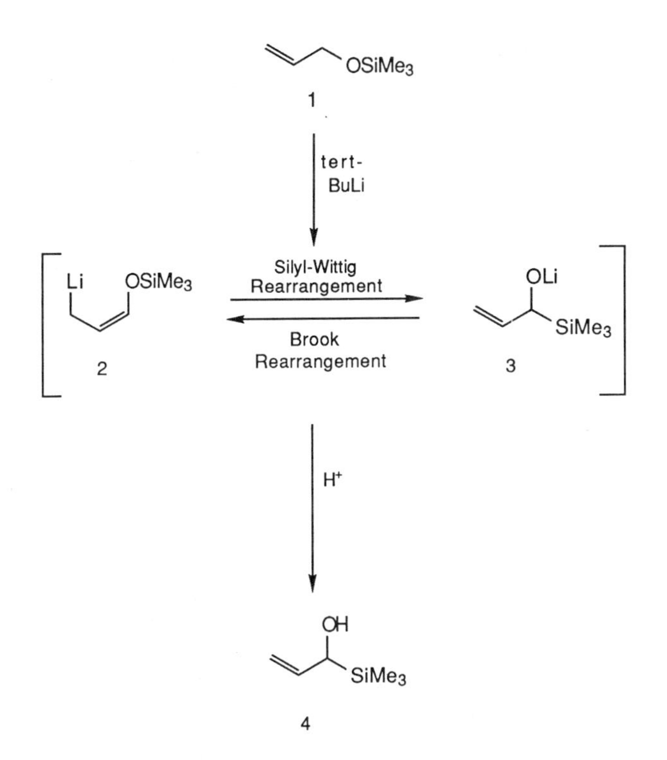

α,β-Unsaturated acylsilanes serve as valuable building blocks for the synthesis of a variety of complex organic compounds. These α,β-unsaturated carbonyl derivatives participate in a number of carbon-carbon bond forming processes including organocuprate conjugate additions,[3] $TiCl_4$-mediated conjugate allylations,[7] Diels-Alder reactions,[3] 1,3-dipolar cycloadditions,[8] and the [3+2] annulation method recently developed in our laboratory.[8] The utility of these reactions is enhanced by the fact that the product acylsilanes are subject to a variety of useful further transformations,[9] including, for example, Brook reactions,[3,10] oxidation to carboxylic acids,[11] and fluoride-promoted conversion to ketones and aldehydes.[11b,12] The present procedure provides a practical method for the preparation of multigram quantities of the simplest α,β-unsaturated acylsilane.

1. Department of Chemistry, Massachusetts Institute of Technology, Cambridge, MA 02139.
2. (a) Watson, S. C.; Eastham, J. F. *J. Organomet. Chem.* **1967**, *9*, 165; (b) Lipton, M. F.; Sorensen, C. M.; Sadler, A. C.; Shapiro, R. H. *J. Organomet. Chem.* **1980**, *186*, 155.
3. Reich, H. J.; Kelly, M. J.; Olson, R. E.; Holtan, R. C. *Tetrahedron* **1983**, *39*, 949.
4. Danheiser, R. L.; Fink, D. M.; Okano, K.; Tsai, Y.-M.; Szczepanski, S. W. *J. Org. Chem.* **1985**, *50*, 5393.
5. For examples of the metalation of allyl silyl ethers, see (a) Still, W. C.; Macdonald, T. L. *J. Am. Chem. Soc.* **1974**, *96* 5561; (b) Still, W. C. *J. Org. Chem.* **1976**, *41*, 3063; (c) Still, W. C.; Macdonald, T. L. *J. Org. Chem.* **1976**, *41*, 3620; (d) Hosomi, A.; Hashimoto, H.; Sakurai, H. *J. Org. Chem.* **1978**, *43*, 2551; (e) Hosomi, A.; Hashimoto, H.; Sakurai, H. *J. Organomet. Chem.* **1979**, *175*, C1; (f) Lau, P. W. K.; Chan, T. H. *J. Organomet. Chem.* **1979**, *179*, C24.
6. Mancuso, A. J.; Huang, S.-L.; Swern, D. *J. Org. Chem.* **1978**, *43*, 2480.
7. Danheiser, R. L.; Fink, D. M. *Tetrahedron Lett.* **1985**, *26*, 2509.
8. Danheiser, R. L.; Fink, D. M. *Tetrahedron Lett.* **1985**, *26*, 2513.
9. For recent reviews on the chemistry of acylsilanes, see (a) Fleming, I. In "Comprehensive Organic Chemistry", Barton, D. H. R.; Ollis, W. D., Eds.; Pergamon Press: Oxford, 1979; Vol. 3, pp 647-653; (b) Magnus, P. D.; Sarkar, T.; Djuric, S. In "Comprehensive Organometallic Chemistry", Wilkinson, G.; Stone, F. G. A.; Abel, E. W., Eds.; Pergamon Press: Oxford, 1982; Vol. 7, pp. 631-639.

10. (a) Brook, A. G. *Acc. Chem. Res.* **1974**, *7*, 77; (b) Brook, A. G.; Bassindale, A. R. In "Rearrangements in Ground and Excited States", de Mayo, P., Ed.; Academic Press: New York, 1980; Vol. 2, pp. 149-227.

11. (a) Zweifel, G.; Backlund, S. J. *J. Am. Chem. Soc.* **1977**, *99*, 3184; (b) Miller, J. A.; Zweifel, G. *J. Am. Chem. Soc.* **1981**, *103*, 6217.

12. (a) Sato, T.; Arai, M.; Kuwajima, I. *J. Am. Chem. Soc.* **1977**, *99*, 5827; (b) Degl'Innocenti, A.; Pike, S.; Walton, D. R. M.; Seconi, G.; Ricci, A.; Fiorenza, M. *J. Chem. Soc., Chem. Commun.* **1980**, 1201; (c) Schinzer, D.; Heathcock, C. H. *Tetrahedron Lett.* **1981**, *22*, 1881; (d) Ricci, A.; Degl'Innocenti, A.; Chimichi, S.; Fiorenza, M.; Rossini, G.; Bestmann, H. J. *J. Org. Chem.* **1985**, *50*, 130.

Appendix

Chemical Abstracts Nomenclature (Collective Index Number); (Registry Number)

(1-Oxo-2-propenyl)trimethylsilane: Silane, trimethyl(1-oxo-2-propenyl)- (9); (51023-60-0)

(1-Hydroxy-2-propenyl)trimethylsilane: 2-Propen-1-ol, 1-(trimethylsilyl)- (11); (95061-68-0)

Allyl alcohol (8); 2-Propen-1-ol (9); (107-18-6)

Chlorotrimethylsilane: Silane, chlorotrimethyl- (8,9); (75-77-4)

Oxalyl chloride (8); Ethanedioyl dichloride (9); (79-37-8)

3-tert-Butyl-4-hydroxy-5-methylphenyl sulfide: o-Cresol, 4,4'-thiobis[6-tert-butyl- (8); Phenyl, 4,4'-thiobis[2-(1,1-dimethylethyl)-6-methyl- (9); (96-66-2)

ETHYL (E,Z)-2,4-DECADIENOATE

(2,4-Decadienoic acid, ethyl ester, (E,Z)-)

A. $C_5H_{11}CH(OH)C{\equiv}CH + CH_3C(OEt)_3 \xrightarrow{CH_3CH_2CO_2H} (H)(C_5H_{11})C{=}C{=}C(CH_2CO_2Et)(H)$ (**1**) $+ 2\,EtOH$

B. $(H)(C_5H_{11})C{=}C{=}C(CH_2CO_2Et)(H)$ (**1**) $\xrightarrow{Al_2O_3}$ $(H)(C_5H_{11})C{=}C(H){-}C(H){=}C(H)(CO_2Et)$ (**2**)

Submitted by S. Tsuboi, T. Masuda, S. Mimura, and A. Takeda.[1]

Checked by Mark A. Henderson and Clayton H. Heathcock.

Caution! See benzene warning, Org. Synth. **1978**, *58*, 168.

A. Ethyl 3,4-decadienoate (1).[2] A 300-mL, round-bottomed flask equipped with a reflux condenser is charged with 12.1 g (0.096 mol) of 1-octyn-3-ol (Note 1), 100 g (0.616 mol) of triethyl orthoacetate (Note 2), and 0.24 g (3.2 mmol) of propionic acid. The solution is heated at 140-150°C in an oil bath. Every 2 hr, the ethanol produced is removed under reduced pressure with a rotary evaporator, and then 10 g (0.062 mol) of triethyl orthoacetate and 0.024 g (0.32 mmol) of propionic acid are added. The mixture is heated until the starting material is consumed (6-8 hr) (Note 3). Excess triethyl orthoacetate is removed under reduced pressure (Note 4). The residue is distilled under reduced pressure to give 15.4-17.2 g (82-91%) of 1 (Note 5) as a clean oil, bp 80-85°C (0.3 mm).

B. Ethyl (E,Z)-2,4-decadienoate (**2**). A dry, 500-mL, round-bottomed flask is charged with 50 g of aluminum oxide (Note 6) and heated at 200°C for 2 hr under reduced pressure (0.05 mm). The flask is fitted with a reflux condenser connected to a nitrogen line and a heavy magnetic stirring bar is added (Note 7); the flask is flushed with nitrogen. With positive nitrogen pressure, the flask is charged wih 200 mL of benzene (Note 8) and 15.4-17.2 g (78-88 mmol) of allenic ester **1**. The mixture is heated at reflux temperature with vigorous stirring for 5 hr. The aluminum oxide is removed by filtration with suction through a sintered-glass funnel of medium porosity, and thoroughly washed with 100 mL of ethyl acetate (Note 9). The combined filtrate is concentrated with a rotary evaporator to afford 11.6-13.6 g (75-88%) of nearly pure **2** as a clean oil (Notes 10 and 11), bp 83-88°C (0.1 mm).

2. Notes.

1. 1-Octyn-3-ol was used as supplied by Aldrich Chemical Company, Inc. (96% purity).

2. Triethyl orthoacetate was used as supplied by Aldrich Chemical Company, Inc. (97% purity) or by Tokyo Kasei Kogyo Co., Ltd. (98% purity).

3. The checkers removed the ethanol with a rotary evaporator and replaced it with fresh triethyl orthoacetate (10 g) and propionic acid (0.024 g) after 2, 4, and 6 hr.

4. The checkers removed excess triethyl orthoacetate under reduced pressure (0.05 mm) overnight. The recovered material may be easily purified by distillation and reused.

5. The product might contain trace amounts of triethyl orthoacetate, but it can be used for the next step since aluminum oxide adsorbs triethyl orthoacetate. The product[2] is characterized by IR (neat) cm^{-1}: 1970 and 1740; ^1H NMR (CCl_4) δ: 0.92 (t, 3 H, $CH_3(CH_2)_4$), 1.29 (t, 3 H, OCH_2CH_3), 1.32 (m, 6 H, $CH_3(CH_2)_3$), 1.91 (m, 2 H, $CH_2CH=$), 2.90 (m, 2 H, CH_2CO_2Et), 4.08 (q, 2 H, $CO_2CH_2CH_3$), 5.09 (m, 2 H, CH=C=CH).

6. Weakly basic alumina[3] (200-300 mesh) for column chromatography is used. The checkers used Alumina Woelm-B, Akt. 1.

7. Efficient stirring is essential to the success of this reaction.

8. Aprotic solvents such as xylene, chlorobenzene, and toluene can be used instead of benzene. If the boiling point of the product is close to that of the solvent, the mixture of aluminum oxide and the allene may be heated to distil at ca. 150°C under an atmosphere of nitrogen.

9. The checkers found that the use of more than 100 mL of ethyl acetate gives a less pure final product.

10. Gas chromatographic analysis (capillary column coated with thermon-1000, 30 m, 140°C) indicated that the product was ca. 93% pure. Impurities consisted of the 2E,4E isomer (5%) and an unidentified compound (2%).

11. The proton magnetic resonance spectrum is as follows (CCl_4) δ: 0.90 (t, 3 H, J = 6, $CH_3(CH_2)_4$), 1.28 (t, 3 H, J = 7, $CO_2CH_2CH_3$), 1.40 (m, 6 H, $CH_3(CH_2)_3CH_2$), 1.9-2.6 (m, 2 H, $CH_2CH=CH$), 4.12 (q, 2 H, J = 7, $CO_2CH_2CH_3$), 5.4-6.3 (m, 3 H, $CH=CH-CH=CHCO_2Et$), 7.46 (dd, 1 H, J = 10 and 15, $CH=CHCO_2Et$); n_D^{25} 1.4895.

3. Discussion

Ethyl (E,Z)-2,4-decadienoate has been prepared in several ways: (a) the addition of lithium di-(Z)-1-heptenylcuprate to ethyl propiolate[4] (90% yield, 95% purity; 27-32% overall yield based on (Z)-1-bromoheptene),[5] (b) the reaction of 1-heptenylmagnesium bromide with ethyl (E)-β-(N,N-diethylamino)acrylate (32% yield, 89% purity),[6] (c) the Wittig reaction of hexyltriphenylphosphonium bromide with ethyl (E)-4-oxo-2-butenoate (68% yield, 85% purity).[7] These known methods involving the use of organometallic reagents need anhydrous conditions at low temperatures (-8° to -40°C). The separation of triphenylphosphine oxide from the reaction mixture in a Wittig reaction is occasionally not easy.

The present procedure offers an experimentally simple and less expensive preparation of ethyl (E,Z)-2,4-decadienoate under essentially neutral conditions. It allows large scale preparation since the starting materials are not sensitive to air or moisture. In addition, the reaction proceeds stereoselectively, and the yields of product are generally high. Several examples are listed in Table 1 to show the scope of the method.

There are many compounds containing a conjugated (E,Z)-diene structure in naturally occurring compounds such as flavors,[3,8] insect pheromones,[3,9] and leukotrienes.[10] The present procedure has been used for the syntheses of bombykol,[3] megatomoic acid,[11] (±)-leukotriene A_4 methyl ester,[12] and (E,Z)-2,4-dienamides.[13]

TABLE I

Rearrangement of β–Allenic Esters to (E,Z)-2,4-Dienoic Esters with Alumina[2]

Starting Material	Product	Yield (%)[a]	Purity (%
 CH_3 CO_2Me	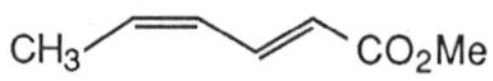 CH_3 CO_2Me	57	90
 C_2H_5 CO_2Me	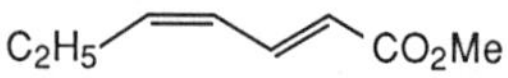 C_2H_5 CO_2Me	82	96
 C_3H_7 CO_2Me	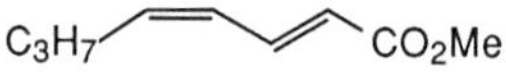 C_3H_7 CO_2Me	80	96
 C_3H_7 CO_2Et	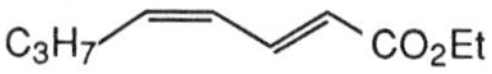 C_3H_7 CO_2Et	69	93
 C_4H_9 CO_2Me	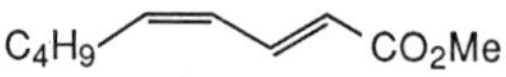 C_4H_9 CO_2Me	87	93
 C_6H_{13} CO_2Me	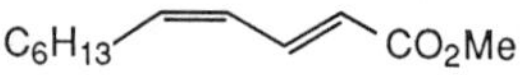 C_6H_{13} CO_2Me	82	99
 C_8H_{17} CO_2Me	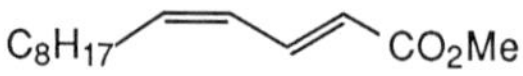 C_8H_{17} CO_2Me	87	91
 C_9H_{19} CO_2Me	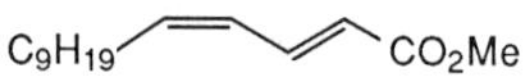 C_9H_{19} CO_2Me	70	96

[a]Isolated yield

[b]Purity determined by gas chromatography

1. Department of Synthetic Chemistry, School of Engineering, Okayama University, Okayama 700, Japan.

2. Morisaki, M.; Bloch, K. *Biochemistry* **1972**, *11*, 309.

3. Tsuboi, S.; Masuda, T.; Takeda, A. *J. Org. Chem.* **1982**, *47*, 4478.

4. Näf, F.; Degen, P. *Helv. Chim. Acta* **1971**, *54*, 1939.

5. Bachman, G. B. *J. Am. Chem. Soc.* **1933**, *55*, 4279.

6. Näf, F.; Decorzant, R. *Helv. Chim. Acta* **1974**, *57*, 1309.

7. Baumann, M.; Hoffmann, W. *Synthesis* **1977**, 681.

8. For example, see: Ohloff, G.; Pawlak, M. *Helv. Chim. Acta* **1973**, *56*, 1176; Buttery, R. G.; Seifert, R. M.; Guadagni, D. G.; Ling, L. C. J. *Arg. Food Chem.* **1971**, *19*, 524.

9. For example, see: Rossi, R. *Synthesis* **1977**, 817.

10. For example, see: Rokach, J.; Adams, J. *Acc. Chem. Res.* **1985**, *18*, 87; Hashimoto, S.; Toda, M. *Yuki Gosei Kagaku Kyokaishi* **1983**, *41*, 221; *Chem. Abstr.* **1983**, *98*, 191841f.

11. Tsuboi, S.; Masuda, T.; Takeda, A. *Bull. Chem. Soc. Jpn.* **1983**, *56*, 3521.

12. Tsuboi, S.; Masuda, T.; Takeda, A. *Chem. Lett.* **1983**, 1829.

13. Tsuboi, S.; Nooda, Y.; Takeda, A. *J. Org. Chem.* **1984**, *49*, 1204.

Appendix

Chemical Abstracts Nomenclature (Collective Index Number);

(Registry Number)

Ethyl (E,Z)-2,4-decadienoate: 2,4-Decadienoic acid, ethyl ester, (E,Z)- (8,9); (3025-30-7)

Ethyl 3,4-Decadienoate: 3,4-Decadienoic acid, ethyl ester (9); (36186-28-4)

1-Octyn-3-ol (8,9); (818-72-4)

Triethyl orthoacetate: Orthoacetic acid, triethyl ester (8); Ethane, 1,1,1-triethoxy- (9); (78-39-7)

α-UNSUBSTITUTED γ,δ-UNSATURATED ALDEHYDES BY CLAISEN REARRANGEMENT: 3-PHENYL-4-PENTENAL

(Benzenepropanal, β-ethenyl-)

A. Me_3N + $HC\equiv CCO_2Et$ $\xrightarrow[CH_2Cl_2]{H_2O}$ $Me_3\overset{+}{N}CH=CHCO_2^-$

B. $Me_3\overset{+}{N}CH=CHCO_2^-$ + $HOCH_2CH=CHPh$ $\xrightarrow[2.\ H_3O^+]{1.\ NaH/THF}$ $PhCH=CHCH_2OCH=CHCO_2H$

C. $PhCH=CHCH_2OCH=CHCO_2H$ $\xrightarrow[-CO_2]{160\text{-}180°C}$ $CH_2=CHCH(Ph)CH_2CHO$

Submitted by Dennis E. Vogel and George H. Büchi.[1]

Checked by Tadahito Nobori and Ryoji Noyori.

1. Procedure

A. (E)-(Carboxyvinyl)trimethylammonium betaine. A 1-L, three-necked, round-bottomed flask is equipped with a mechanical stirrer, dropping funnel, and thermometer. The flask is charged with 25.0 g (0.255 mol) of ethyl propiolate (Note 1), 14 mL of dichloromethane and 440 mL of water. The mixture is cooled to 5°C (Note 2) and 90 mL (0.35 mol) of an aqueous 25% solution of trimethylamine (Note 3) is added under vigorous stirring over a period of 30 min. The reaction temperature remains between 0 and 5°C during the addition and then is allowed to warm to 25°C for 3 hr. The dichloromethane layer is separated and the aqueous layer is washed three times with 100-mL portions of dichloromethane. The aqueous layer is placed in a 1-L, round-bottomed flask and concentrated at 15 mm pressure with a rotary evaporator equipped with a dry ice condenser. The flask is heated to approximately 45°C. When the residue gives the appearance of a wet solid, it is treated with 150 mL of dioxane and concentrated as described above. Dioxane treatment followed by concentration is repeated three more times with 150-mL portions of dioxane (Note 4). The yellow solid residue is triturated with acetonitrile (Note 5) until a white solid is obtained. The solid is collected and dried at 0.1 mm, 25°C for 14 hr to give 25.0-26.9 g (76-82% yield) of (E)-(carboxyvinyl)trimethylammonium betaine, mp 176-177°C (dec) (Note 6).

B. (E)-3-[(E)-3-Phenyl-2-propenoxy]acrylic acid. An oven dried, 1-L, three-necked, round-bottomed flask equipped with a mechanical stirrer, dropping funnel and reflux condenser is purged with argon and charged with 8.20 g (0.171 mol) of 50% sodium hydride in oil (Note 7) and 90 mL of anhydrous tetrahydrofuran (Note 8). To this mixture is added a solution of 19.0 g (0.143 mol) of cinnamyl alcohol (Note 9) in 180 mL of anhydrous tetrahydrofuran. The mixture is stirred for 30 min at which point 25.0 g (0.194 mol) of (E)-(carboxyvinyl)trimethylammonium betaine is added and the reaction mixture is heated at a gentle reflux for 15 hr. The cooled reaction mixture is slowly added to a mixture of 600 mL of water and 220 mL of a saturated aqueous solution of sodium chloride (Note 10). The residual material is removed with wet ether (Note 11) and added to the aqueous solution. The aqueous layer is washed three times with 450 mL of ether and acidified with approximately 21 mL of concentrated hydrochloric acid to pH 1. This mixture is extracted three times with 700 mL of ether and the ether layer is dried with 20 g of anhydrous magnesium sulfate for 30 min. The mixture is filtered and the filtrate is concentrated, first with a rotary evaporator and then at 0.1 mm for 24 hr at 25°C to give crude (E)-3-[(E)-3-phenyl-2-propenoxy]acrylic acid. This material can be used in Part C without further purification. However, purification by trituration with approximately 30 mL of anhydrous ether gives 24.7-25.0 g (85% yield) of white crystalline product, mp 140-141.5°C (dec) (Note 12).

C. 3-Phenyl-4-pentenal. In a 100-mL, round-bottomed flask equipped with a magnetic stirring bar is placed 20.1 g (0.098 mol) of (E)-3-[(E)-3-phenyl-2-propenoxy]acrylic acid from B (Note 13). The flask is fitted with a distillation head for vacuum distillation and heated at 0.1 mm pressure. The oil bath temperature is maintained between 160-165°C while the mixture is stirred, until all the material melts. Once the initial reaction is under control the oil bath is slowly heated to 180°C. The product is collected in a receiver flask cooled with a dry ice-acetone bath to give 13.3-14.3 g (84-91% yield) of 3-phenyl-4-pentenal, bp 114-115°C (15 mm) (Note 14).

2. Notes

1. Ethyl propiolate, purchased from Aldrich Chemical Company, Inc., is freshly distilled prior to use.

2. An ice-acetone bath is used to cool the reaction mixture.

3. An aqueous 25% solution of trimethylamine available from Aldrich Chemical Company, Inc. is used directly.

4. This procedure helps to concentrate further the product by forming an azeotrope with the water.

5. Continued trituration with small portions of acetonitrile (e.g., eight times with 50-mL portions) eventually removes all traces of the yellow-colored impurity as well as any residual water.

6. The product exhibits the following spectral properties: IR (KBr) cm^{-1}: 1665, 1600, 1360; ^1H NMR (D_2O, DSS ref) δ: 3.3 (br s, 9 H, $(CH_3)_3N$), 6.3 (br d, 1 H, J = 13, vinyl CH), 6.8 (br d, 1 H, J = 13, vinyl CH).

7. A suspension of 50% sodium hydride in mineral oil, purchased from Alfa Products, Morton/Thiokol Inc. or Nakarai Chemicals, Ltd., is used directly.

8. Anhydrous tetrahydrofuran is obtained by distillation from benzophenone ketyl prior to use.

9. Cinnamyl alcohol purchased from Aldrich Chemical Company, Inc. is freshly distilled before use. The checkers purchased it from Nakarai Chemicals, Ltd.

10. *Caution: Residual sodium hydride is present in the reaction mixture; however, it can be quenched safely by strict adherence to the procedure described.*

11. The use of wet ether has two purposes. Not only does the ether help wash residue product from the flask, but ether which has not been carefully dried contains traces of water which allows for the safe quenching of the last traces of sodium hydride.

12. The product exhibits the following spectral properties: IR (KBr) cm^{-1}: 1680, 1615, 1600; ^1H NMR ($CDCl_3$ 60 MHz) δ: 4.5 (d, 2 H, J = 5.5, $OCH_2CH=$), 5.2 (d, 1 H, J = 12, $CH=CHCO_2H$), 6.5 (m, 2 H, PhCH=CH-), 7.2 (s, 5 H, Ph-), 7.5 (d, 1 H, J = 12, $O-CH=CHCO_2H$), 12.0 (s, 1 H, CO_2H).

13. The material obtained from trituration of the crude product in Part B can also be used at this point and gives similar results.

14. The product exhibits the following spectral properties: IR (neat) cm^{-1}: 2840, 2740, 1720, 1640, 1600; ^{1}H NMR ($CDCl_3$, 60 MHz) δ: 2.9 (d,d 2 H, J = 2, 4, CH_2-CHO), 4.0 (q, 1 H, J = 7, Ph-CH), 5.0 (d, 1 H, J = 7, vinyl), 5.3 (s, 1 H, vinyl), 6.0 (m, 1 H, vinyl), 7.2 (s, 5 H, Ph), 9.5 (t, 1 H, J = 2, CHO).

3. Discussion

The procedure described[2] illustrates a new general synthetic method for the preparation of (E)-3-allyloxyacrylic acids and their conversion to α-unsubstituted γ,δ-unsaturated aldehydes by subsequent Claisen rearrangement-decarboxylation. Such aldehydes are traditionally prepared by Claisen rearrangements of allyl vinyl ethers.[3] Allyl vinyl ethers are typically prepared by either mercury-catalyzed vinyl ether exchange with allylic alcohols or acid-catalyzed vinylation of allylic alcohols with acetals. The basic conditions required for alkoxide addition to the betaine to produce carboxyvinyl allyl ethers, as described in this report, nicely complements these two methods. In addition, this Claisen rearrangement is an experimentally very simple procedure, since sealed tube and other high pressure vessels are not required. The allyloxyacrylic acids are heated neat (in most cases a small amount of hydroquinone is added) and, by adjusting the pressure at which the reaction is performed, the aldehyde products distill from the reaction mixture in analytically pure form.

(E)-(Carboxyvinyl)trimethylammonium betaine is prepared by a modification of the procedure of McCulloch and McInnes.[4] They also reported the addition of simple alkoxides to this betaine, and that deuterium exchange is not observed when this reaction is performed with the deuterated betaine. We have also observed that replacement of the betaine with propiolic acid leads to the formation of 3-alkoxyacrylic acids in significantly lower yields.[2] These observations are best accounted for by an addition-elimination process.

This procedure provides a variety of allyloxyacrylic acids; however, it is sensitive to steric hindrance. Tertiary allylic alcohols do not add to the betaine and sterically hindered secondary alcohols add with decreasing facility. Table I indicates the scope of this reaction.

1. Department of Chemistry, Massachusetts Institute of Technology, Cambridge, MA 02139.
2. Büchi, G.; Vogel, D. E. *J. Org. Chem.* **1983**, *48*, 5406.
3. Saucy, G.; Marbet, R. *Helv. Chim. Acta* **1967**, *50*, 2091; Watanabe, W. H.; Conlon, L. E. *J. Am. Chem. Soc.* **1957**, *79*, 2828; Church, R. F.; Ireland, R. E.; Marshall, J. A. *J. Org. Chem.* **1966**, *31*, 2526; Thomas, A. F. *J. Am. Chem. Soc.* **1969**, *91*, 3281; Dauben, W. G.; Dietsche, T. J. *J. Org. Chem.* **1972**, *37*, 1212; Cookson, R. C.; Rogers, N. R. *J. Chem. Soc., Perkin Trans. 1* **1973**, 2741; Cookson, R. C. ; Rogers, N. R. *J. Chem. Soc., Chem. Commun.* **1972**, 248; Cookson, R. C.; Hughes, N. W. *J. Chem. Soc., Perkins Tran. 1* **1973**, 2738.
4. McCulloch, A. W.; McInnes, A. G. *Can. J. Chem.* **1974**, *52*, 3569.

TABLE I

PREPARATION AND REARRANGEMENT OF 3-ALLYLOXYACRYLIC ACIDS

tarting Material	Yield of Adduct, %	Product	Yield, %
OH	89	CHO	68
OH	84	CHO	78
Ph OH	92	CHO Ph	quant.
OH	62	CHO	quant.
OH	96	CHO	quant.
OH	53,77	CHO	quant.
O OH	82	CHO O	79
		O CHO	16
OH	55	CHO	82

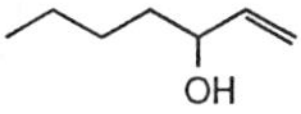

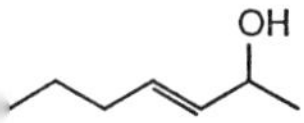

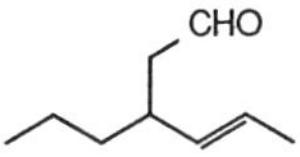

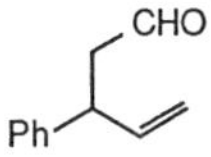

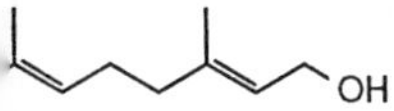

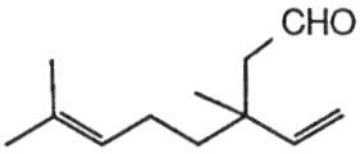

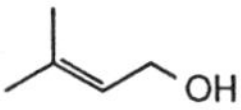

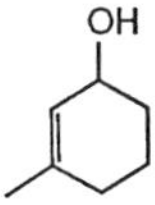

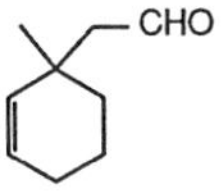

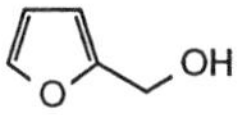

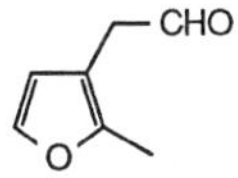

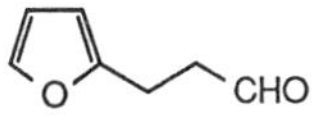

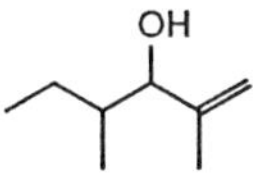

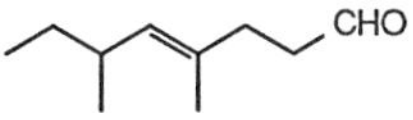

Appendix

Chemical Abstracts Nomenclature (Collective Index Number); (Registry Number)

3-Phenyl-4-pentenal: Benzenepropanal, β-ethenyl- (9); (939-21-9)

(E)-(Carboxyvinyl)trimethylammonium betaine: Ethenaminium, 2-carboxy-N,N,N-trimethyl-, hydroxide, inner salt, (E)- (9); (54299-83-1)

Ethyl propiolate: Propiolic acid, ethyl ester (8); 2-Propynoic acid, ethyl ester (9); (623-47-2)

(E)-3-[(E)-3-Phenyl-2-propenoxy]acrylic acid: 2-Propenoic acid, 3-[(3-phenyl-2-propenyl)oxy]-, (E,E)- (10); (88083-18-5)

Cinnamyl alcohol (8); 2-Propen-1-ol, 3-phenyl- (9); (104-54-1)

APROTIC DOUBLE MICHAEL ADDITION:

PREPARATION OF 1,3-DIMETHYL-5-OXOBICYCLO[2.2.2]OCTANE-2-CARBOXYLIC ACID

(Bicyclo[2.2.2]octane-2-carboxylic acid, 1,3-Dimethyl-5-oxo-)

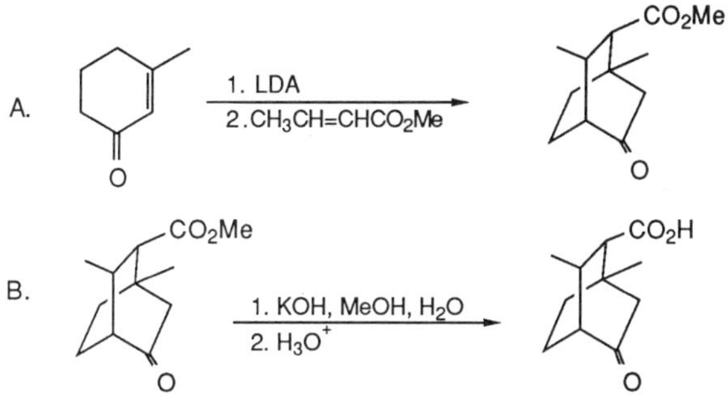

Submitted by Dietrich Spitzner and Anita Engler.[1]

Checked by Michael P. Trova, Mary A. Kinsella, and Leo A. Paquette.

1. Procedure

A. Methyl 1,3-dimethyl-5-oxobicyclo[2.2.2]octane-2-carboxylate.[2] An oven-dried, 250-mL, round-bottomed flask equipped with a stirring bar and a rubber septum is charged with 100 mL of dry tetrahydrofuran and 5.56 g (55 mmol) of anhydrous diisopropylamine. The flask is flushed with argon via a needle inlet-outlet and cooled to -78°C with a dry ice-isopropyl alcohol bath (Notes 1-3). To the stirred solution is added dropwise with a syringe 30 mL (54 mmol) of a 1.8 M solution of butyllithium in hexane (Note 4) to form lithium diisopropylamide, followed after 30 min by a solution of 5.50 g (50 mmol) of 3-methyl-2-cyclohexen-1-one (Note 5) in 60 mL of dry tetrahydrofuran via a flex-needle over a period of 15 min. Stirring and cooling is continued for an additional 30 min. To the resulting yellow solution of the lithium dienolate, 10.0 g (0.1 mol) of methyl (E)-crotonate (Note 6) is added with a syringe within 2 min. The cooling bath is removed and the reaction mixture is allowed to warm to room temperature (Note 7). Stirring is continued at room temperature for 2 hr. The reaction mixture is quenched by adding 1 N hydrochloric acid until the mixture turns acidic. Extraction with three 80-mL portions of dichloromethane followed by evaporation of the solvent yields a light yellow oil which is taken up in 100 mL of diethyl ether. This solution is filtered through 100 g of silica gel to remove polymers and water. Elution with diethyl ether and evaporation of the solvent gives a yellow oil which is distilled in a Kugelrohr distillation apparatus (Note 8) under reduced pressure. After a small forerun (ca. 0.5 g of unreacted 3-methyl-2-cyclohexen-1-one) at 50°C, 0.05 mm, the main fraction is collected at 110-120°C (oven temperature), 0.05 mm to give 8.25-9.43 g (78-90%) of a colorless oil, which solidifies on standing in the freezer. One recrystallization from cold pentane (approximately 10 mL of pentane per 8 g of ester mixture) gives 6.0 g of product as white crystals, mp 37°C (Note 9), of approximately 97% isomeric purity (Note 10).

B. *1,3-Dimethyl-5-oxobicyclo[2.2.2]octane-2-carboxylic acid.* A mixture of 11.2 g (53.4 mmol) of the foregoing crude ester in 40 mL of methanol and 8.0 g (143 mmol) of potassium hydroxide in 16 mL of water is refluxed under argon until the ester is no longer present when monitored by TLC (Note 7). This takes about 1 day. Methanol is distilled off with a rotary evaporator, and the remaining dark solution is extracted with ether (2 x 50 mL), acidified to pH 1 with dilute sulfuric acid and extracted with dichloromethane (4 x 50 mL). The organic layer is filtered through 100 g of silica gel and eluted with ether to remove most of the dark impurities. Concentration under reduced pressure gives 10.0 g of acid mixture. Distillation in a Kugelrohr apparatus at 180°C, 0.03 mm and one recrystallization from ether-pentane gives 7.0 g (67%) of pure bicyclic acid (isomeric purity greater than 98%) as white crystals, mp 130-131°C (Note 11).

2. Notes

1. All glassware, syringes and flex-needles were baked in an oven at 120°C overnight and assembled while hot.

2. Tetrahydrofuran was purified by passing it through activated (12 hr at 450°C) neutral aluminum oxide purchased from ICN and distilling it fresh from lithium aluminum hydride.

3. Diisopropylamine was distilled from calcium hydride prior to use.

4. Butyllithium in hexane was purchased from Metallgesellschaft AG, Frankfurt, Germany and standardized by titration with diphenylacetic acid.[3]

5. 3-Methyl-2-cyclohexen-1-one was purchased from Aldrich-Europe, but is easily prepared from ethyl acetoacetate and paraformaldehyde.[4]

6. Methyl crotonate may polymerize to some extent under these conditions. An excess is used in order to insure complete formation of the product. Unreacted methyl crotonate is easily removed by distillation.

7. The reaction was monitored by TLC (silica gel 60PF254, Merck, Darmstadt, Germany; 1:1 diethyl ether:pentane as the mobile phase, 2,4-dinitrophenylhydrazine as revealing reagent) and by GLC (N_2, 3% SE30 rubber on Volaspher A2, Merck, Darmstadt, Germany, 140°C isotherm). The Michael reaction is very slow at -78°C, and the optimum temperature depends upon the acceptor (Table I).

8. A Büchi rotary evaporator was used.

9. The spectra are as follows: IR (neat) cm^{-1}: 1730 (ester, ketone); EI-GCMS (70 eV): m/e = 210 (M^+, 5%), 110 (100), 95 (30); ^{1}H-NMR (250 MHz, $CDCl_3$, TMS) δ: 0.94 (s, Me, 3 H), 1.10 (d, 3 H, J = 7), 1.30-2.35 (m, 8 H), 2.75 (dd, 1 H, J = 3 and 19), 3.67 (s, OMe).

10. The oily product contains approximately 8% of the exo isomer (estimated by ^{1}H NMR on the basis of the ester methyl at 3.70 ppm (major) and at 3.67 ppm (minor)).

11. An additional 1.2 g (9.5%) of pure acid may be recovered from the mother liquor. The spectra are as follows: ^{13}C NMR (62.88 MHz, $CDCl_3$) δ: 17.3 (t), 17.4 (q), 23.7 (q), 31.8 (d), 33.9 (t), 35.7 (s), 44.9 (t), 47.9 (d), 50.9 (q), 54.5 (d), 174.8 (s), 214.1 (s); ^{1}H NMR (250 MHz, $CDCl_3$) δ: 1.05 (s, 3 H), 1.15 (d, 3 H, J = 6.8), 1.35-2.15 (m, 8 H), 2.32 (m, 1 H), 2.81 (dd, 1 H, J = 19, 3), 11.2 (broad s, 1 H); EI-MS (70 eV): m/z = 196 (M^+, 40%), 178 (5), 110 (100), 95(45); IR (CH_2Cl_2) cm^{-1}: 3480(m), 2920(s), 1725(s), 1705(s).

3. Discussion

The aprotic double Michael addition was discovered by R. A. Lee[5] and used[6-10] to synthesize functionalized bicyclo[2.2.2]octanes which may serve as starting materials in natural products syntheses (Table I). These bicyclo[2.2.2]octanes can also be obtained by a Diels-Alder cycloaddition of 2-trimethylsiloxy-substituted cyclohexadienes and dienophiles:[11]

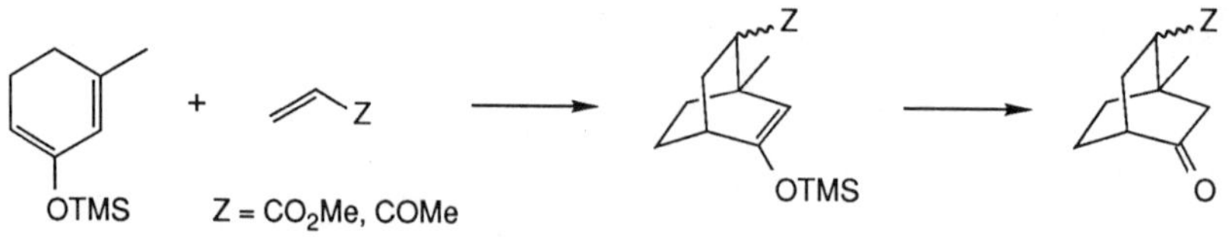

But there are many cases known where the (4+2) cycloaddition fails even with siloxy-activated dienes, e.g., methyl (E)-crotonate does not react with diene **1** at normal pressure and elevated temperature (110°C), whereas the aprotic double Michael addition does give the desired bicyclo[2.2.2]octane in high yield. This reaction gives mainly (92%) bicyclic esters with the endo configuration.

TABLE I

EXAMPLES OF CARBOCYCLIC ESTERS PREPARED BY THE APROTIC DOUBLE MICHAEL ADDITION

Dienolate	Acceptor	Product	Yield(%)	Lit.
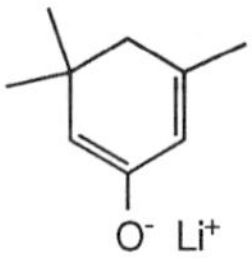 $O^- Li^+$	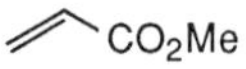 CO_2Me	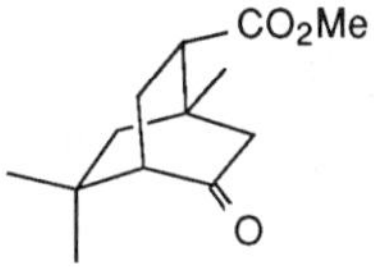 CO_2Me O	98	4
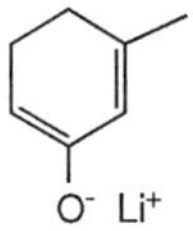 $O^- Li^+$	 CO_2Me	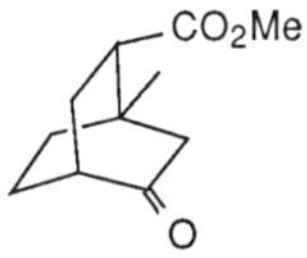 CO_2Me O	90	5
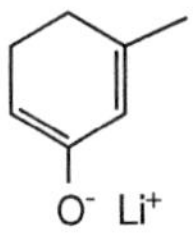 $O^- Li^+$	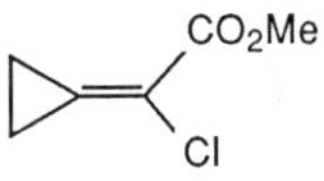 CO_2Me Cl	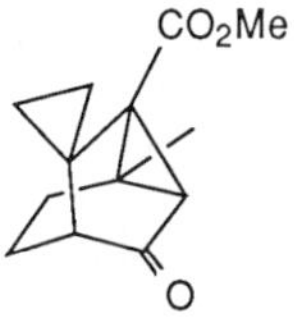 CO_2Me O	64	8
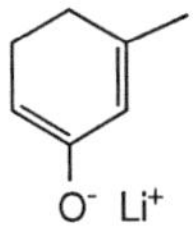 $O^- Li^+$	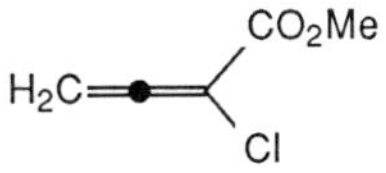 H_2C CO_2Me Cl	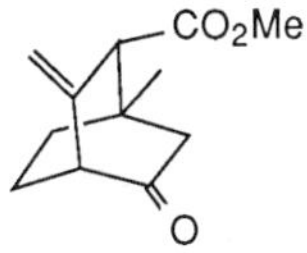 CO_2Me O	75	9

1. Institut für Chemie, University Hohenheim, Garbenstr. 30, D-7000 Stuttgart 70, Germany.
2. Hagiwara, H.; Uda, H.; Kodama, T. *J. Chem. Soc., Perkin Trans, I* **1980**, 963.
3. Kofron, W. G.; Baclawski, L. M. *J. Org. Chem.* **1976**, *41*, 1879.
4. Cronyn, M. W.; Riesser, G. H. *J. Am. Chem. Soc.* **1953**, *75*, 1664.
5. Lee, R. A. *Tetrahedron Lett.* **1973**, 3333.
6. Spitzner, D. *Tetrahedron Lett.* **1978**, 3349.
7. White, K. B.; Reusch, W. *Tetrahedron* **1978**, *34*, 2439.
8. Roberts, M. R.; Schlessinger, R. H. *J. Am. Chem. Soc.* **1981**, *103*, 724.
9. Spitzner, D.; Engler, A.; Liese, T.; Splettstosser, G.; de Meijere, A. *Angew. Chem. Suppl.* **1982**, 1722.
10. Spitzner, D. *Angew. Chem. Intern. Ed. Engl.* **1982**, *21*, 636.
11. Brownbridge, P. *Synthesis* **1983**, 85; Rubottom, G. M.; Krueger, D. S. *Tetrahedron Lett.* **1977**, 611; Jung, M. E.; McCombs, C. A. *Tetrahedron Lett.* **1976**, 2935.

Appendix

Chemical Abstracts Nomenclature (Collective Index Number); (Registry Number)

3-Methyl-2-cyclohexen-1-one: 2-Cyclohexen-1-one, 3-methyl- (8,9); (1193-18-6)

(E)-Methyl crotonate: Crotonic acid, methyl ester, (E)- (8); 2-Butenoic acid, methyl ester, (E)- (9); (623-43-8)

COPPER-CATALYZED CONJUGATE ADDITION OF A ZINC HOMOENOLATE: ETHYL 3-[3-(TRIMETHYLSILYLOXY)CYCLOHEX-2-ENYL)]PROPIONATE

(2-Cyclohexene-1-propanoic acid, 3-[(trimethylsilyl)oxy]-, ethyl ester)

$$2\ \text{(OEt)(OSiMe}_3\text{)cyclopropane} + ZnCl_2 \xrightarrow{Et_2O} Zn(CH_2CH_2CO_2Et)_2 + 2\ Me_3SiCl$$

$$Zn(CH_2CH_2CO_2Et)_2 + \text{cyclohexenone} \xrightarrow[Me_3SiCl]{CuBr \cdot Me_2S} \text{(OSiMe}_3\text{)cyclohexenyl–CH}_2\text{CH}_2\text{CO}_2\text{Et}$$

Submitted by Eiichi Nakamura and Isao Kuwajima.[1]

Checked by Tina M. Kravetz, Daniel Cheney, and Leo A. Paquette.

1. Procedure

In a tared 1-L, three-necked flask, two necks of which are covered with rubber septa, and the other connected to a nitrogen/vacuum source, is placed 17.2 g of zinc chloride (Note 1). The flask is evacuated to approximately 2 mm and heated with a burner with swirling until practically all of the salt melts. The flask is cooled and filled with nitrogen. The dried salt weighs 16.4-17 g (ca. 0.12 mol) (Note 2). An efficient magnetic stirring bar and a Dimroth condenser in place of a rubber septum are set in position, and the flask is again flushed with nitrogen. Ether (300 mL) (Note 3) is introduced via the septum, and stirring is initiated and maintained throughout the reaction. The mixture is refluxed gently for 1 hr to aid dissolution of the solid salt (Note 4). The flask is cooled, and 1-trimethylsilyloxy-1-ethoxycyclopropane (41.80 g, 0.24 mol) (Note 5) is introduced with the aid of a hypodermic syringe during 5 min. The cloudy mixture is stirred at room temperature for 1 hr; the more dense lower layer may mostly have disappeared at this point. The mixture is refluxed for 30 min to complete homoenolate formation. The clear, colorless solution of the zinc homoenolate and chlorotrimethylsilane is cooled in an ice bath, and cuprous bromide/dimethyl sulfide complex (0.4 g, 2 mmol) (Note 6) is added by removing the septum while nitrogen adequate to exclude air is introduced through the inlet. 2-Cyclohexen-1-one (9.62 g, 0.1 mmol) (Note 7) is introduced via the septum during 1 min, and then hexamethylphosphoric triamide (HMPA) (34.8 mL, 0.2 mol) (Notes 8,9) is added during 5 min. A slightly exothermic reaction occurs initially and the bath is removed after 20 min. After 3 hr at room temperature, 40 g of silica gel (Note 10) and 300 mL of dry hexane (Note 11) are added while the mixture is stirred vigorously for 3 min. The supernatant liquid is decanted, and the residue is suspended in 60 mL of dry ether. Dry hexane (60 mL) is added and the supernatant liquid is decanted. This extractive procedure is repeated once and the combined organic phase is filtered through Celite (Note 12). After concentration with a rotary evaporator, the oily product is distilled under reduced pressure (ca. 2 mm). 1-Trimethylsilyloxy-1-ethoxycyclopropane (8-10 g) is recovered as the first fraction (bp 26°C/2.3 mm). The majority of the HMPA remaining after workup distils at 80-120°C/2.3 mm. Finally, the desired product (18.9-20.5 g, 70-76%) is obtained as a fraction boiling at 130-132°C/2.3 mm (Note 13).

2. Notes

1. Zinc chloride was purchased from Koso Chemical Company and used as such (cf., *Org. Synth.* **1974**, *54*, 49). Alfa's ultra pure grade reagent resisted complete dissolution and appeared less suitable. The checkers used "Baker Analyzed" reagent zinc chloride with prior drying at 0.3 mm.

2. The amount of zinc chloride may be in slight excess of the theoretical amount (i.e., 0.5 equiv of the cyclopropane).

3. Ether was distilled from sodium benzophenone ketyl immediately before use.

4. A two-layer mixture results.

5. This cyclopropane was prepared according to an *Organic Syntheses* procedure, *Org. Synth.* **1985**, *63*, 147.

6. Cuprous bromide/dimethyl sulfide complex was purchased from Aldrich Chemical Company, Inc. and used as such.

7. Cyclohexenone was purchased from Tokyo Kasei Chemical Company or Aldrich Chemical Company, Inc. and used after simple distillation at reduced pressure.

8. Hexamethylphosphoric triamide (HMPA) was purchased from Tokyo Kasei Chemical Company and used after distillation from calcium hydride under reduced pressure.

9. This step was also checked substituting N,N'-dimethylpropyleneurea (DMPU),[2] supplied by Aldrich Chemical Company, Inc. or Fluka, for HMPA. The yield of final product dropped somewhat to 16.36-17.36 g (60-64%), but otherwise the reaction proceeded as described.

10. Ordinary silica gel (Wakogel C-300, Wako Chemical Company) was used.

11. Hexane was distilled from calcium hydride and stored over a potassium mirror. The checkers stored the redistilled hexane over molecular sieves.

12. Since the product has only moderate hydrolytic stability, the extractive procedure should be carried out rapidly under a flow of nitrogen. Operation in a nitrogen-filled plastic bag may eliminate the possibility of hydrolysis.

13. When the workup is performed as described (Note 11), the product may contain small amounts of residual HMPA and up to 5% (GLC estimation) of the keto ester resulting from hydrolysis of the enol silyl ether. Rf values of the keto ester and the silyl ether on thin layer analysis (Merck silica gel plates coated with a 0.25-mm layer of Kieselgel 60 F_{254}, developed with 30% ethyl acetate in hexane) were 0.4 and 0.8, respectively; gas chromatographic (GLC) analysis (OV-101, capillary glass column of 0.25-mm x 20-m, 120°C) showed retention times of 2.56 and 4.96 min for these two compounds, respectively. GLC analysis also indicated the ratio of the regioisomers of the enol silyl ether as >99:1 (4.96 and 6.13 min, respectively). On a smaller scale where the product can readily be handled on silica gel chromatography, a yield over 85% may be attained. Correct elemental analysis has been obtained for a sample purified by chromatography and distillation. Spectral properties of the product are as follows: ^{1}H NMR (300 MHz, CCl_4) δ: 0.04 (s, 9 H), 1.10 (t, 3 H, J = 7.1), 1.3-2.2 (m, 11 H), 3.93 (q, 2 H, J = 7.1), 4.55 (br s, 1 H); IR (neat film) cm^{-1} 1730 (s), 1655 (s), 1445 (m), 1365 (s), 1245 (s), 1180 (vs), 840 (vs), 745 (s).

3. Discussion

Unlike their enolate counterparts, homoenolates have been underrated because of a prior lack of synthetic accessibility.[3] Many of the previously known homoenolates cyclize readily to the cyclopropanolate tautomer and behave chemically as the latter. 1-Alkoxy-1-silyloxycyclopropanes[4] have provided, for the first time, examples of reactive yet characterizable homoenolates (of alkyl propionates). A titanium homoenolate undergoes 1,2-addition to carbonyl compounds, providing an efficient synthetic route to γ-lactones.[4] The present procedure represents an unique and highly efficient method for the preparation of the zinc homoenolate of an alkyl propionate and illustrates its copper-catalyzed conjugate addition.[5] The reaction consists of two stages; the first part of the present procedure generates a mixture of the zinc homoenolate and chlorotrimethylsilane, from which the homoenolate can be isolated by removal of the volatile material under reduced pressure, and the second part involves the chlorotrimethylsilane-assisted conjugate addition of the transient copper homoenolate. Only one of the propionate moieties on the zinc metal is available for the conjugate addition. The reaction mechanism has already been discussed briefly.[6]

The reaction is applicable to a variety of enones, enals, and acetylenic carbonyl compounds (Table). No 1,2-addition is seen under copper-catalyzed conditions since the zinc homoenolate does not generally undergo a 1,2-addition reaction to carbonyl compounds. The conjugate adduct is useful for organic synthesis as indicated by the scheme on the next page. The enol silyl ether moiety acts either to protect or activate the ketone functionality. The ready hydrolytic generation of the keto ester from the conjugate adduct provides an efficient entry to 6-keto esters. Replacement of the enone with an acyl halide leads to 4-keto esters in high yield,[5] and the palladium catalyzed reaction with aryl and vinyl halides gives 3-aryl- and 3-vinyl propionates.[8] The purified homoenolate undergoes 1,2-addition to aldehydes in the presence of chlorotrimethylsilane.[9]

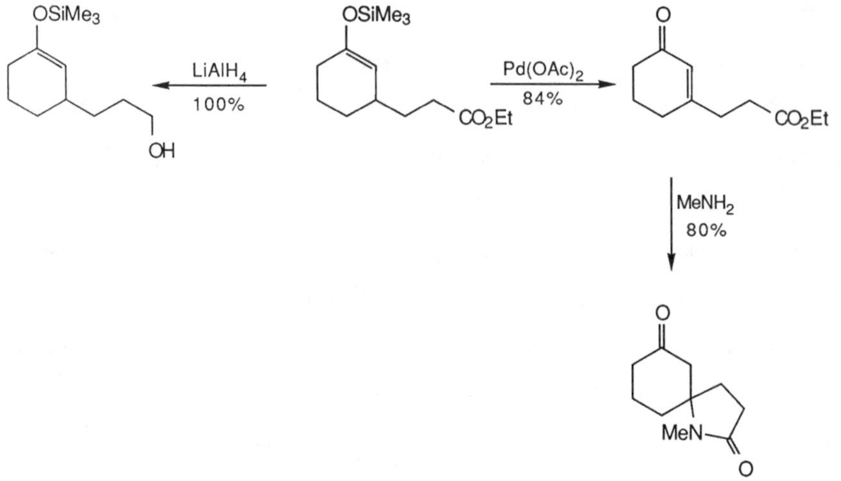

No other synthetic method is known that achieves the equivalent transformation. Rather elaborate procedures using an allylic anion type of the homoenolate "equivalents"[6] or homoenolate radicals[7] have been reported, but their tolerance to the structure of the enone acceptor is much narrower.

1. Department of Chemistry, Tokyo Institute of Technology, Meguro, Tokyo 152, Japan.

2. Mukhopadhyay, T.; Seebach, D. *Helv. Chim. Acta* **1982**, *65*, 385.

3. Werstiuk, N. H. *Tetrahedron* **1983**, *39*, 205.

4. Nakamura, E.; Kuwajima, I. *J. Am. Chem. Soc.* **1977**, *99*, 7360; Nakamura, E.; Kuwajima, I. *J. Am. Chem. Soc.* **1983**, *105*, 651; Nakamura, E.; Kuwajima, I. *J. Am. Chem. Soc.* **1985**, *107*, 2138; Nakamura, E.; Oshino, H.; Kuwajima, I. *J. Am. Chem. Soc.* **1986**, *108*, 3745.

5. Nakamura, E.; Kuwajima, I. *J. Am. Chem. Soc.* **1984**, *106*, 3368; Nakamura, E.; Shimada, J.-i., Kuwajima, I. *Organometallics* **1985**, *4*, 641.

6. Lesur, B.; Toye, J.; Chantrenne, M.; Ghosez, L. *Tetrahedron Lett.* **1979**, 2835; De Lombaert, S.; Lesur, B.; Ghosez, L. *Tetrahedron Lett.* **1982**, *23*, 4251; Evans, D. A.; Baillargeon, D. J.; Nelson, J. V. *J. Am. Chem. Soc.* **1978**, *100*, 2242.

7. Giese, B.; Horler, H. *Tetrahedron Lett.* **1983**, *24*, 3221.

8. Nakamura, E.; Kuwajima, I. *Tetrahedron Lett.* **1986**, *27*, 83.

9. Oshino, H.; Nakamura, E.; Kuwajima, I. *J. Am. Chem. Soc.* **1985**, *50*, 2802.

TABLE

CONJUGATE ADDITION OF HOMOENOLATE OF ESTERS

Acceptor	Product	%Yield
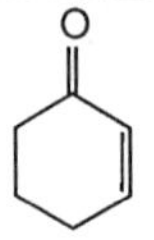 O	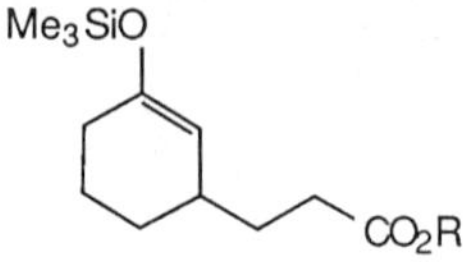 Me_3SiO … CO_2R	R = i-Pr 93 R = Et 76 R = Me 91
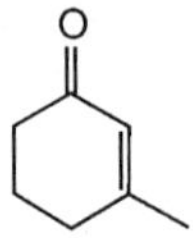 O	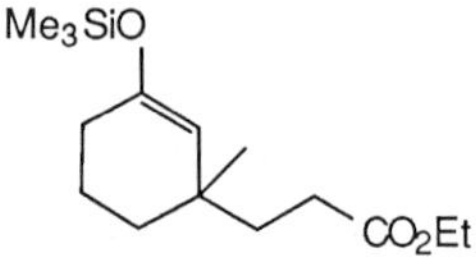 Me_3SiO … CO_2Et	92
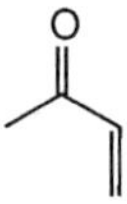 O	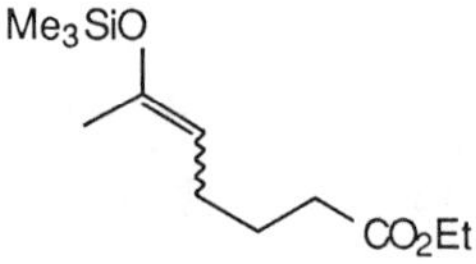 Me_3SiO … CO_2Et	80
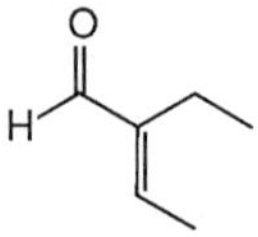 O, H	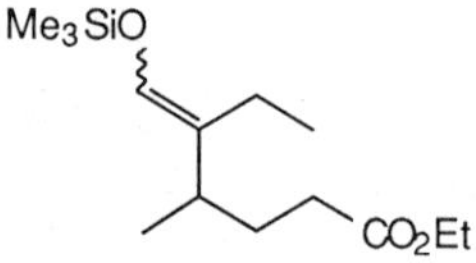 Me_3SiO … CO_2Et	75
 O	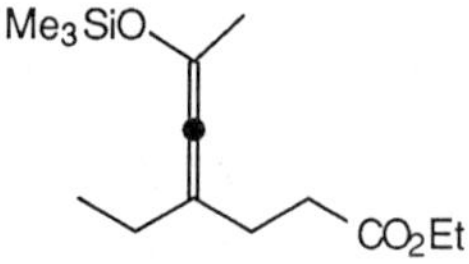 Me_3SiO … CO_2Et	73
 MeO_2C … MeO_2C	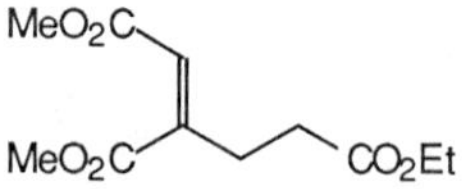 MeO_2C, MeO_2C … CO_2Et	63

Appendix

Chemical Abstracts Nomenclature (Collective Index Number); (Registry Number)

Ethyl 3-[3-(trimethylsilyloxy)cyclohex-2-enyl)]propionate: 2-Cyclohexene-1-propanoic acid, 3-[(trimethylsilyl)oxy]-, ethyl ester (11); (90147-64-1)

Zinc chloride (8,9); (7646-85-7)

1-Trimethylsilyloxy-1-ethoxycyclopropane:

Silane, [(1-ethoxycyclopropyl)oxy]trimethyl- (8,9); (27374-25-0)

Cuprous bromide/dimethyl sulfide: Copper, bromo[thiobis[methane]]- (9); (54678-23-8)

Cyclohexenone: 2-Cyclohexen-1-one (8,9); (930-68-7)

Hexamethylphosphoric triamide: Phosphoric triamide, hexamethyl- (8,9); (680-31-9)

N,N'-Dimethylpropyleneurea: 2(1H)-Pyrimidone, tetrahydro-1,3-dimethyl- (8,9); (7226-23-5)

CONJUGATE ADDITION/CYCLIZATION OF A CYANOCUPRATE:
2-CARBOMETHOXY-3-VINYLCYCLOPENTANONE
(Cyclopentanecarboxylic acid, 2-ethenyl-5-oxo-, methyl ester)

A. 2 $CH_2=CHCO_2Me$ $\xrightarrow[LiBF_4]{Pd(NCMe)_4(BF_4)_2}$ $MeO_2CCH_2CH_2CH=CHCO_2Me$

B. $MeO_2CCH_2CH_2CH=CHCO_2Me$ $\xrightarrow[2)\ H_3O^+]{1)\ Li_2Cu(CN)(CH=CH_2)_2}$ 2-carbomethoxy-3-vinylcyclopentanone (CO_2Me, $CH=CH_2$)

Submitted by William A. Nugent and Frank W. Hobbs, Jr.[1]

Checked by David J. Wustrow and Andrew S. Kende.

1. Procedure

A. Dimethyl (E)-2-hexenedioate.[2] A 100-mL, one-necked, round-bottomed flask is capped by a septum, swept with dry nitrogen and flame-dried. The flask is charged with methyl acrylate (50 mL, 0.55 mol, Note 1), then anhydrous lithium tetrafluoroborate (9 g, 0.096 mol, Note 2), and finally tetrakis(acetonitrile)palladium tetrafluoroborate (1.33 g, 0.003 mmol, Note 3). The mixture is stirred briefly until homogeneous. It is warmed under nitrogen in a 40°C-oil bath for 72 hr (Note 4) and then allowed to cool to room temperature. The mixture is added to saturated aqueous sodium bicarbonate (100 mL) and extracted with ether (3 x 50 mL). The combined ether extracts are dried over anhydrous magnesium sulfate, filtered and concentrated to an oil with a rotary evaporator. The residue is distilled through a 10-cm

Vigreux column to give dimethyl (E)-2-hexenedioate (38.6 g, 81%, Note 5) as a colorless liquid, bp 100°C (1.1 mm).

B. 2-Carbomethoxy-3-vinylcyclopentanone. A 1-L, three-necked, round-bottomed flask is fitted with a 125-mL pressure-equalizing addition funnel capped with a septum, an overhead stirrer and a septum. The apparatus is flame-dried and purged with dry nitrogen. The flask is charged through the addition funnel with tetravinyltin (12.48 g, 0.055 mol, Note 6) and anhydrous ether (250 mL). The solution is cooled to 0°C under nitrogen, and low-halide methyllithium in ether (133 mL, 1.5 M, 0.20 mol, Note 7) is slowly added directly by syringe to the stirred solution. After 15 min, the vinyllithium mixture is cooled in a dry ice-acetone bath to -78°C for 20 min. The septum on one neck is briefly removed, and copper(I) cyanide (9.31 g 0.107 mol, Note 8) is added all at once. The septum is replaced by a low-temperature thermometer in an adapter. The bath and reaction are allowed to warm under nitrogen slowly, with stirring, so that the internal temperature is -30°C after 1 hr (Note 9). The addition funnel is charged with dimethyl (E)-2-hexenedioate (6.89 g, 0.040 mol) and anhydrous ether (16 mL). The contents of the addition funnel are added dropwise over 30 min to the cuprate at -30°C, and stirring is continued under the nitrogen atmosphere for an additional 30 min at that temperature. A mixture of saturated aqueous ammonium chloride (80 mL) and water (80 mL) is added dropwise over 20 min through the addition funnel while the temperature of the system is allowed to rise. After the mixture is stirred for an additional 90 min it is filtered through a medium-porosity glass frit. The flask and filter cake are rinsed with water (2 x 30 mL) and ether (2 x 30 mL). The ether layer is separated and the aqueous layer is further extracted with ether (2 x 75 mL). The combined organic layers are washed with water (25 mL), dried over anhydrous magnesium sulfate and concentrated with a rotary evaporator (Note 10). The residue (Note 11) is distilled through a short-path distillation apparatus to afford 2-carbomethoxy-3-vinylcyclopentanone (5.39 g, 80%, Note 12) as a colorless liquid, bp 65-70°C at 0.4 mm (Note 13).

2. Notes

1. Reagent grade methyl acrylate from Fisher Scientific Company, containing p-methoxyphenol as inhibitor, was used as received.

2. Anhydrous 98% pure lithium tetrafluoroborate ($LiBF_4$) from Alfa Products, Morton/Thiokol, Inc. was used as received.

3. The palladium complex was purchased from Strem Chemical Company and was used as received. Alternatively, material prepared from palladium sponge and nitrosonium tetrafluoroborate in acetonitrile[3] worked equally well.

4. The greyish precipitate which begins to appear after ca. 40 hr is the 1:1 adduct of the product with lithium tetrafluoroborate.

5. Submitters find that the product typically contains 95% of 2-hexenedioates as measured by capillary column GLC (30 m DB17 column, 120°C isothermal). Retention times for the isomeric hexenedioates were (Z)-2 (2.44 min), (Z)-3 (2.75 min), (E)-3 (3.08 min), (E)-2 (3.44 min). TLC (30:70 ethyl acetate/hexane, UV) for some runs shows, in addition to the product at R_f = 0.36, a weak spot due to an intensely UV-active impurity at R_f = 0.41. The spectra are as follows: ^1H NMR ($CDCl_3$) δ: 2.45-2.58 (m, 4 H), 3.69 (s, 3 H), 3.73 (s, 3 H), 5.87 (d, 1 H, J = 16), 6.96 (dt, 1 H, J = 16, 6); IR (CCl_4) cm^{-1}: (C=O) 1743 s, 1730 s, (C=C) 1661 m.

6. The submitters obtained tetravinyltin from Columbia Organic Chemicals Company; it was used as received. The checkers obtained it from K&K Laboratories, ICN Biomedicals Inc., Plainview, N.Y. It may also be synthesized by literature methods.[4]

7. Low-halide methyllithium in ether from Alfa Products, Morton/Thiokol, Inc. or Aldrich Chemical Company, Inc. was used as received. A single run using methyllithium/lithium bromide complex gave a significantly reduced yield (53%). Use of commercial vinyllithium in tetrahydrofuran gave a product contaminated with starting dimer, requiring chromatographic purification.

8. Copper(I) cyanide from Alfa Products, Morton/Thiokol, Inc. was used as received. *Caution! Copper(I) cyanide is severely toxic.* Care should be taken not to expose cyanide-containing wastes to strong acid thus liberating hydrogen cyanide. Prior to disposal, insoluble wastes should be treated overnight with a strong alkaline solution containing calcium hypochlorite.

9. If the internal temperature is allowed to rise too quickly, rapid exothermic cuprate formation can occur with resultant decomposition of the reagent.

10. Tetramethyltin (bp 78°C) is a potentially hazardous side-product of this reaction. This work-up should therefore be done with gloves in a well-ventilated hood. Most of the tetramethyltin ends up in the condensate from the rotary evaporator; the condensate should be disposed of by incineration.

11. In two cases submitters have observed that the residue separated into two layers. The upper layer consists of a heavy oil apparently because of incomplete washing of the lithium suspension used in manufacturing methyllithium. When this happens it is necessary to remove the oil with a pipette prior to distillation. Failure to do so gives a product which appears pure by TLC, but which is substantially impure according to elemental analysis (1% high in carbon).

12. Submitters find that the product is homogeneous by TLC (30:70 ethyl acetate/hexane, I_2, R_f = 0.43). Capillary column GLC analysis (30 m DB17 column, 120°C isothermal) is complicated by some thermal decarboxylation on the column. However, using a clean injection port liner and 180°C injection port, 95% of the product is eluted as a single, somewhat broad peak at 3.0 min retention time. The spectra are as follows: ^1H NMR ($CDCl_3$) δ: 1.72 (m, 1 H), 2.1-2.6 (m, 3 H), 3.05 (d, 1 H, J = 11), 3.1-3.3 (m, 1 H), 3.75 (s, 3 H), 5.09 (d, 1 H, J = 11), 5.16 (d, 1 H, J = 17), 5.75-5.85 (m, 1 H); IR (CCl_4) cm^{-1}): (C=O) 1762 s, 1735 s, 1662 m, 1618 m; (C=C) 1644 w.

13. The submitters have carried out these steps on twice the scale given here. On that scale their yields for step A were 91-93%, for step B, 77-85%.

The checkers found that the diastereomeric purity of the product was much greater than 90% based upon its 300 MHz ^1H and fully decoupled ^{13}C NMR spectra. Based on the proton-proton coupling constant (J = 11), trans geometry has been assigned.

3. Discussion

This procedure illustrates a general route to the 3-substituted 2-carbomethoxycyclopentanones, which are versatile intermediates for the preparation of a variety of cyclopentanoid products. For example, the product of this procedure, 2-carbomethoxy-3-vinylcyclopentanone, has been utilized in the synthesis of methyl dihydrojasmonate[5] and 18-hydroxyestrone.[6] This conjugate addition/cyclization approach (utilizing "Gilman reagents" prepared from copper(I) iodide) has been applied[7] to the synthesis of the methyl, butyl, sec-butyl, neopentyl, and phenyl substituted 2-carbomethoxycyclopentanones. The present procedure takes advantage of the greater stability of higher order cyanocuprates[8] ("Lipshutz reagents") to overcome the moderate yield of the vinyl analogue due to cuprate decomposition as reported in earlier studies.[7] With either the Gilman or Lipshutz reagents, Michael addition to dimethyl (E)-2-hexenedioate produces an enolate which undergoes Dieckmann cyclization faster than proton transfer. Therefore, no 4-substituted cyclopentanones are formed. This approach has now been extended to the synthesis of the corresponding cyclopentenones by using dimethyl 2-hexynedioate as the Michael acceptor.[9]

Alternatively, 3-substituted 2-carbomethoxycyclopentanones have been prepared by Michael addition to 2-carbomethoxycyclopentenone.[10-12] However, this Michael acceptor is unstable, difficult to prepare, and polymerizes in the presence of many nucleophiles.[11] A longer synthesis of 2-carbomethoxy-3-vinylcyclopentanone has been reported.[5] The general route to 2-carbomethoxy-3-vinylcyclopentanones developed by Trost[13] has the advantage of producing these compounds in optically active form.

Several catalyst systems have been described for the tail-to-tail dimerization of methyl acrylate.[14-19] Advantages of the dimerization procedure described here are its mild conditions, efficient use of the catalyst, and high selectivity for the Δ^2 isomer. Tetrakis(acetonitrile)palladium tetrafluoroborate, $Pd(NCMe)_4(BF_4)_2$, also efficiently catalyzes dimerization of ethyl acrylate and methyl methacrylate. The presence of lithium tetrafluoroborate in the reaction mixture increases the rate of the reaction and prolongs catalyst life. Dimerization of methyl acrylate can be effected without lithium tetrafluoroborate if the reaction is performed in nitromethane.

1. Central Research and Development Department, E. I. du Pont de Nemours and Company, Experimental Station, Wilmington, DE 19898.
2. Commercial use of the acrylate dimerization procedure is restricted under U.S. Patent No. 4 451 665 (E. I du Pont de Nemours and Company).
3. Wayland, B. B.; Schramm, R. F. *Inorg. Chem.* **1969**, *8*, 971-976.
4. (a) Seyferth, D.; Stone, F. G. A. *J. Am. Chem. Soc.* **1957**, *79*, 515-517; (b) Rosenberg, S. D.; Gibbons, A. J., Jr.; Ramsden, H. E. *J. Am. Chem. Soc.* **1957**, *79*, 2137-2140.
5. Tsuji, J.; Kobayashi, Y.; Kataoka, H.; Takahashi, T. *Tetrahedron Lett.* **1980**, *21*, 1475-1478.
6. Tsuji, J.; Okumoto, H.; Kobayashi, Y.; Takahashi, T. *Tetrahedron Lett.* **1981**, *22*, 1357-1358.
7. Nugent, W. A.; Hobbs, F. W., Jr. *J. Org. Chem.* **1983**, *48*, 5364-5366.
8. Lipshutz, B. H.; Wilhelm, R. S.; Kozlowski, J. *Tetrahedron Lett.* **1982**, *23*, 3755-3758.
9. Crimmins, M. T.; Mascarella, S. W.; DeLoach, J. A. *J. Org. Chem.* **1984**, *49*, 3033-3035.
10. Marx, J. N.; Minaskanian, G. *Tetrahedron Lett.* **1979**, 4175-4178.
11. Marx, J. N.; Cox, J. H.; Norman, L. R. *J. Org. Chem.* **1972**, *37*, 4489-4491.
12. Marx, J. N.; Minaskanian, G. *J. Org. Chem.* **1982**, *47*, 3306-3310.
13. Trost, B. M.; Runge, T. A. *J. Am. Chem. Soc.* **1981**, *103*, 7559-7572.
14. Alderson, T.; Jenner, E. L.; Lindsey, R. V., Jr. *J. Am. Chem. Soc.* **1965**, *87*, 5638-5645.
15. Barlow, M. G.; Bryant, M. J.; Haszeldine, R. N.; Mackie, A. G. *J. Organomet. Chem.* **1970**, *21*, 215-226.
16. Pracejus, H.; Krause, H.-J.; Oehme, G. *Z. Chem.* **1980**, *20*, 24.
17. Oehme, G.; Pracejus, H. *Tetrahedron Lett.* **1979**, 343-344.

18. Grenouillet, P.; Neibecker, D.; Tkatchenko, I. *Organometallics* **1984**, *3*, 1130-1132.

19. Nugent, W. A.; McKinney, R. J. *J. Mol. Catal.* **1985**, *29*, 65-76.

Appendix

Chemical Abstracts Nomenclature (Collective Index Number); (Registry Number)

2-Carbomethoxy-3-vinylcyclopentanone: Cyclopentanecarboxylic acid, 2-ethenyl-5-oxo-, methyl ester (10); (75351-19-8)

Dimethyl (E)-2-hexenedioate: 2-Hexenedioic acid, dimethyl ester, (E)- (10); (70353-99-0)

Methyl acrylate: 2-Propenoic acid, methyl ester (9); (96-33-3)

Tetrakis(acetonitrile)palladium tetrafluoroborate: Palladium(2+), tetrakis(acetonitrile)-, bis[tetrafluoroborate (1-)] (8,9); (21797-13-7)

PALLADIUM-CATALYZED SYNTHESIS OF CONJUGATED DIENES: (5Z,7E)-5,7-HEXADECADIENE

A. $C_8H_{17}C{\equiv}CH$ + $(i\text{-}C_4H_9)_2AlH$ ⟶ (E)-$C_8H_{17}CH{=}CHAl(i\text{-}C_4H_9)_2$

B. (E)-$C_8H_{17}CH{=}CHAl(i\text{-}C_4H_9)_2$ + (Z)-$ICH{=}CHC_4H_9$ —(1 mol % $Pd(PPh_3)_4$, $ZnCl_2$)⟶ $C_8H_{17}CH{=}CHCH{=}CHC_4H_9$

Submitted by Ei-ichi Negishi, Tamotsu Takahashi, and Shigeru Baba.[1]

Checked by Masako Ohta and Ryoji Noyori.

1. Procedure

Caution! Organoaluminum compounds are pyrophoric. They must be kept and used with caution under a nitrogen atmosphere.

A. (E)-1-Decenyldiisobutylalane. An oven-dried, 300-mL, two-necked, round-bottomed flask equipped with a magnetic stirring bar, a rubber septum inlet, and an outlet connected to a mercury bubbler is flushed with nitrogen, immersed in a water bath kept at room temperature, and charged with 22.6 mL (125 mmol) of 1-decyne (Note 1) and 80 mL of hexane (Note 2). To this flask is added dropwise, with stirring, 22.3 mL (125 mmol) of diisobutylaluminum hydride (Note 3) using a syringe (Note 4). After the addition has been completed, the water bath is replaced with an oil bath kept at 50-60°C, and the reaction mixture is stirred for 6 hr at this temperature (Note 5).

B. (5Z,7E)-5,7-Hexadecadiene. To a mixture of 21.0 g (100 mmol) of (Z)-1-hexenyl iodide (Note 6), 13.6 g (100 mmol) of zinc chloride (Note 7), and 1.15 g (1 mmol) of tetrakis(triphenylphosphine)palladium (Note 8) in 100 mL of tetrahydrofuran (Note 9) is added the solution of (E)-1-decenyldiisobutylalane prepared above, while the reaction temperature is controlled with a water bath at room temperature. After the reaction mixture is stirred for 6 hr at room temperature, it is slowly transferred into a mixture of 300 mL of ice-cooled 3 N hydrochloric acid and 100 mL of pentane via a double-ended needle under a positive pressure of nitrogen. The organic layer is separated, and the aqueous layer is extracted twice with 100 mL of pentane. The combined organic layer is washed with 100 mL of water, followed by 100 mL of saturated aqueous sodium bicarbonate, and then dried over anhydrous magnesium sulfate. After filtration, the solvent is removed using a rotary evaporator. Hydroquinone (30 mg) is added and the residue is distilled (Note 10) to give 14.3-15.8 g (64-66% based on (Z)-1-hexenyl iodide) of (5Z,7E)-5,7-hexadecadiene as a colorless liquid, bp 116-119°C (1 mm) (Notes 11, 12).

2. Notes

1. The submitters used 1-decyne from Farchan Laboratories, Inc., without further purification. The checkers used the material purchased from Aldrich Chemical Company, Inc. and Tokyo Kasei Kogyo Company.

2. Hexane available from Fisher Scientific Company was purified by distillation from sodium. The checkers used the solvent from Wako Pure Chemical Industries, Ltd. after distillation from sodium and benzophenone or calcium hydride under argon.

3. The submitters used neat diisobutylaluminum hydride obtained from Ethyl Corporation. The checkers used the material from Aldrich Chemical Company, Inc.

4. The submitters used a 50-mL syringe with a Luer-lock and a long ($\geq$ 6 inch) 18-gauge needle, the plunger of which was lightly greased with a silicone grease. For a larger scale operation, it is advisable to use a septum-capped graduated cylinder and a double-ended needle in place of a syringe.

5. GLC analysis of a small aliquot after iodinolysis with iodine dissolved in tetrahydrofuran indicated the formation of (E)-1-decenyldiisobutylalane (ca. 80%), 1-decynyldiisobutylalane (7-8%), and 1-decene (5%), together with 1-decyne (ca. 10%).

6. (Z)-1-Hexenyl iodide was prepared by treating acetylene with lithium dibutylcuprate followed by iodine according to an *Organic Syntheses* procedure.[2]

7. Zinc chloride available from Mallinckrodt, Inc., was flame-dried under a slow stream of nitrogen. The checkers used the material from Wako Pure Chemical Industries, Ltd., after heating under vacuum (1 mm).

8. Tetrakis(triphenylphosphine)palladium was prepared according to an Inorganic Syntheses procedure.[3] The submitters used a freshly prepared, shiny yellow crystalline sample of the palladium complex. On standing for an extended period of time (> a few weeks), its color gradually darkens. The checkers used tetrakis(triphenylphosphine)palladium purchased from Aldrich Chemical Company, Inc.

9. Tetrahydrofuran available from Fisher Scientific Company or Wako Pure Chemical Industries, Ltd. was distilled from sodium and benzophenone.

10. Hydroquinone was added to avoid polymerization of the diene product.

11. Gas chromatographic examination of the reaction mixture using a 2-ft column of 20% SE-30 on Chromosorb W with undecane as an internal standard (200°C) indicates that (5Z,7E)-5,7-hexadecadiene is formed in 86-90% yield, based on (Z)-1-hexenyl iodide. The product obtained by this procedure shows the following properties: n_D^{23} 1.4662; IR (neat) cm^{-1}: 1630 (w), 1370 (w), 978 (m), 943 (m), 720 (m); ^1H NMR [CCl_4, $(CH_3)_4Si$] δ: 0.8-1.0 (m, 6 H), 1.15-1.5 (m, 16 H), 1.9-2.3 (m, 4 H) m, 5.15 (dt, 1 H, J = 10, 8), 5.61 (dt, 1 H, J = 14, 7), 5.92 (dd, 1 H, J = 10, 11), 6.30 (dd, 1 H, J = 11, 14).

12. (5E,7E)-5,7-Hexadecadiene can be prepared in the same manner as described except that (E)-1-hexenyl iodide is used in place of its (Z) isomer. The (E) iodide is obtainable by treating (E)-1-hexenyldiisobutylalane, prepared from 1-hexyne and diisobutylaluminum hydride, with iodine in tetrahydrofuran.[4] The (E,E) diene prepared by this procedure shows the following properties: n_D^{23} 1.4671; IR (neat) cm^{-1}: 1370 (w), 982 (m), 722 (m); ^1H NMR [CCl_4, $(CH_3)_4Si$] δ: 0.89 (t, 6 H, J = 6), 1.15-1.7 (m, 16 H), 1.8-2.2 (m, 4H), 5.2-6.15 (m, 4 H): ^{13}C NMR [$CDCl_3$, $(CH_3)_4Si$] δ: 13.93, 14.06, 22.34, 22.76, 29.40, 29.62, 31.16, 31.78, 32.03, 32.37, 32.72, 130.59, 132.19, 132.26.

3. Discussion

The procedure described here is based on two reports by Negishi and his co-workers.[5,6] The original procedure[5] did not use zinc chloride as the second catalyst and required ca. 5-mol % of $Pd(PPh_3)_4$. The preparation of

(E)-1-alkenylalane via hydroalumination was first reported by Wilke and Müller.[7] The procedure used here is essentially that which was described by Zweifel.[8]

The first highly stereoselective and satisfactory syntheses of conjugated dienes of general applicability are those based on organoboron chemistry reported by Negishi for (E,E) dienes[9] and (E,Z) dienes[10] as well as by Zweifel for (Z,Z) dienes.[11]

The method described here represents the first highly selective and general cross-coupling procedure for preparing conjugated dienes.[5,6] Subsequently, several variations of the above-described method have been published by Negishi and others. Some salient features of these investigations are as follows. First, palladium-phosphine complexes, such as $Pd(PPh_3)_4$, are preferred to nickel complexes,[5] which tend to reduce stereoselectivity. Second, all three possible stereoisomers may be prepared using this methodology. Third, in addition to Al, Zr,[12] Mg,[13] B,[14] and Cu[15] have been shown to participate in Pd-catalyzed alkenyl-alkenyl cross coupling. Fourth, the reaction is markedly accelerated by the addition of zinc and cadmium halides, such as $ZnCl_2$, $ZnBr_2$, and $CdCl_2$,[6] at least in cases where organometals containing Al,[6] Zr,[6] and Cu[15] are used. This and other observations suggest that alkenylzinc derivatives are probably the most reactive organometals in this reaction,[6] although no systematic comparisons have so far been made. Otherwise, the choice of metal depends on the stereochemistry of the desired product, the required chemoselectivity, and other factors. For example, (E)-alkenyl metals are most directly available via hydroalumination,[8] carboalumination,[16] hydroboration,[17] or hydrozirconation[18] of alkynes. In cases where the use of such derivatives is desirable, Al, B, or Zr should be considered first. On the other hand, (Z)-

alkenyl metals are often readily available via carbocupration[19] of alkynes, and they may be used directly or via formation of the corresponding (Z)-alkenyl iodide.

Together with the organoboron procedures mentioned above, the Pd-catalyzed cross coupling procedures have been applied to the synthesis of a wide variety of insect pheromones, terpenoids, and carotenoids.

1. Department of Chemistry, Purdue University, West Lafayette, IN 47907.

2. Alexakis, A.; Cahiez, G.; Normant; J. F. *Org. Synth.* **1984**, *62*, 1.
3. Coulson, D. R. *Inorg. Synth.* **1972**, 13, 121.
4. Zweifel, G.; Whitney, C. C. *J. Am. Chem. Soc.* **1967**, *89*, 2753.
5. Baba, S.; Negishi, E. *J. Am. Chem. Soc.* **1976**, *98*, 6729.
6. Negishi, E.; Okukado, N.; King, A. O.; Van Horn, D. E.; Spiegel, B. I. *J. Am. Chem. Soc.* **1978**, *100*, 2254.
7. Wilke, G.; Müller, H. *Chem. Ber.* **1956**, *89*, 444.
8. Zweifel, G.; Miller, J. A. *Org. React.* **1984**, *32*, 375.
9. Negishi, E.; Yoshida, T. *J. Chem. Soc., Chem. Commun.* **1973**, 606.
10. Negishi, E.; Lew, G.; Yoshida, T. *J. Chem. Soc., Chem. Commun.* **1973**, 874.
11. Zweifel, G.; Polston, N. L. *J. Am. Chem. Soc.* **1970**, *92*, 4068.
12. Okukado, N.; Van Horn, D. E.; Klima, W. L.; Negishi, E. *Tetrahedron Lett.* **1978**, 1027.
13. (a) Dang, H. P.; Linstrumelle, G. *Tetrahedron Lett.* **1978**, 191; (b) Ratovelomanana, V.; Linstrumelle, G. *Tetrahedron Lett.* **1981**, *22*, 315.
14. Miyaura, N.; Suginome, H.; Suzuki, A. *Tetrahedron Lett.* **1981**, *22*, 127.
15. Jabri, N.; Alexakis, A.; Normant, J. F. *Tetrahedron Lett.* **1981**, *22*, 959; **1982**, *23*, 1589.

16. (a) Van Horn, D. E.; Negishi, E. *J. Am. Chem. Soc.* **1978**, *100*, 2252; (b) For a review, see Negishi, E. *Pure Appl. Chem.* **1981**, *53* 2333.

17. Brown, H. C. "Hydroboration", Benjamin: New York, 1962.

18. Schwartz, J. *J. Organometal Chem. Library* **1976**, *1*, 461.

19. Normant, J. F.; Alexakis, A. *Synthesis* **1981**, 841.

Appendix

Chemical Abstracts Nomenclature (Collective Index Number); (Registry Number)

1-Decyne (8,9); (764-93-2)

Diisobutylaluminum hydride: Aluminum, hydrodiisobutyl- (8); Aluminum, hydrobis(2-methylpropyl)- (9); (1191-15-7)

(Z)-1-Hexenyl iodide: 1-Hexene, 1-iodo-, (Z)- (8,9); (16538-47-9)

(E)-1-Hexenyl iodide: 1-Hexene, 1-iodo-, (E)- (8,9); (16644-98-7)

Tetrakis(triphenylphosphine)palladium: Palladium, tetrakis(triphenylphosphine)- (8); Palladium, tetrakis(triphenylphosphine)-, (T-4)- (9); (14221-01-3)

(E)-1-Hexenyldiisobutylalane: Aluminum, 1-hexenyldiisobutyl-, (E)- (8); Aluminum, 1-hexenylbis(2-methylpropyl)-, (E)- (9); (20259-40-9)

1-Hexyne (8,9); (693-02-7)

SYNTHESIS OF BIARYLS VIA PALLADIUM-CATALYZED

CROSS COUPLING: 2-METHYL-4'-NITROBIPHENYL

(1,1'-Biphenyl, 2-methyl-4'-nitro-)

A. o-iodotoluene → 1) tert-BuLi, 2) $ZnCl_2$ → o-tolylzinc chloride (CH_3, ZnCl)

B. o-tolylzinc chloride + Br–C_6H_4–NO_2 → $Pd(PPh_3)_4$ → 2-methyl-4'-nitrobiphenyl

Submitted by Ei-ichi Negishi, Tamotsu Takahashi, and Anthony O. King.[1]

Checked by Koji Kawai and Ryoji Noyori.

1. Procedure

A. o-Tolylzinc chloride. An oven-dried, 500-mL, three-necked, round-bottomed flask equipped with a magnetic stirring bar, dropping funnel, rubber septum inlet, and an outlet connected to a mercury bubbler is flushed with nitrogen, immersed in a dry-ice bath kept at -78°C, and charged with 26.2 g (120 mmol) of o-iodotoluene (Note 1) and 60 mL of ether (Note 2). To this solution is added dropwise, with stirring, 154 mL (1.56 M, 240 mmol) of a hexane solution of tert-butyllithium (Note 3). After the reaction mixture is stirred for 1 hr at -78°C, it is warmed to room temperature, stirred for 1 hr and concentrated under diminished pressure using a water aspirator (ca. 15 mm) until most of the volatile solvents are evaporated. To this concentrate is added 80 mL of tetrahydrofuran (THF) (Notes 4 and 5). The mixture is, in turn, added to 16.3 g (120 mmol) of dry zinc chloride (Note 6) and 60 mL of tetrahydrofuran placed in a similarly-equipped, 500-mL flask using a 16-gauge double-ended needle under a slight positive pressure of nitrogen; the resulting mixture is stirred for 1 hr at room temperature.

B. 2-Methyl-4'-nitrobiphenyl. To a mixture of 1.16 g (1 mmol) of tetrakis(triphenylphosphine)palladium (Note 7), 100 mL of tetrahydrofuran and 20.2 g (100 mmol) of 1-bromo-4-nitrobenzene (Note 8) in a 500-mL flask, set up as described above and immersed in a water bath, is added the o-tolylzinc chloride solution prepared above. The reaction mixture is stirred for 6 hr at room temperature and poured onto a mixture of 100 mL of ether and 300 mL of ice-cold 3 N hydrochloric acid. The organic layer is separated, and the aqueous layer is extracted with two 100-mL portions of ether. The combined organic layer is washed with saturated aqueous sodium bicarbonate and dried over anhydrous magnesium sulfate. After filtration, the solvent is removed using a rotary evaporator to give a light brown solid. The solid is recrystallized from 300 mL of hexane to yield 16.0 g of yellow crystals. The second crop from 50 mL of hexane is 2.4 g. The combined product is recrystallized from 100 mL of ethanol to afford 16.0 g of light yellow needles. Crystallization of the mother liquor from 25 mL of ethanol gives a second crop of 0.5 g. The total yield of 2-methyl-4'-nitrobiphenyl is 16.5 g (78% based on 1-bromo-4-nitrobenzene) (Note 9).

2. Notes

1. The submitters used o-iodotoluene from Aldrich Chemical Company, Inc. The checkers purchased it from Wako Chemical Industries, Ltd.

2. Ether available from Fisher Scientific Company or Sanraku Company was distilled from sodium and benzophenone.

3. The submitters used tert-butyllithium from Aldrich Chemical Company, Inc. after titration by the method of Watson and Eastham.[2a] The checkers titrated it by the method of Lipton.[2b]

4. Tetrahydrofuran from Fisher Scientific Company or Kishida Chemical Company was distilled from sodium and benzophenone.

5. Although somewhat more cumbersome, the following more economical procedure using lithium metal may also be used to generate o-tolyllithium. To 1.7 g (240 mg-atom) of freshly cut lithium in 35 mL of ether at 0°C is added dropwise 20.5 g (120 mmol) of o-bromotoluene in 25 mL of ether. After formation of o-tolyllithium, it is diluted with 60 mL of tetrahydrofuran before use.

6. Zinc chloride, available from Mallinckrodt, Inc., was flame-dried under a slow stream of nitrogen in the reaction flask. The checkers used zinc chloride from Wako Pure Chemical Industries, Inc. after fusion by flame-drying under reduced pressure for 20 min.

7. Tetrakis(triphenylphosphine)palladium was prepared according to the procedure of Coulson.[3]

8. The submitters used 1-bromo-4-nitrobenzene from Aldrich Chemical Company, Inc. without further purification. The checkers purchased it from Tokyo Kasei Kogyo Company.

9. Gas chromatographic examination of another reaction mixture run on a 10-mmol scale with undecane as an internal standard indicates that 2-methyl-4'-nitrobiphenyl is formed in 90% yield based on 1-bromo-4-nitrobenzene. The product obtained by this procedure shows the following properties: mp 99-101°C (lit.,[4] mp 103-105°C); IR (neat) cm^{-1}: 1600 (s), 1510 (s), 1480 (s), 1340 (s),

858 (s), 778 (s), 756 (s), 730 (s), 702 (s); ^{1}H NMR (90 MHz, $CDCl_3$) δ: 2.26 (s, 3 H), 7.15-7.40 (m, 4 H), 7.48 (d, 2 H, J = 8.5), 8.27 (d, 2 H, J = 8.5); ^{13}C NMR (22.5 MHz, $CDCl_3$) δ: 20.30, 123.42, 126.16, 128.50, 129.42, 130.11, 130.75, 135.09, 139.67, 147.00, 148.85.

3. Discussion

The procedure described above is based on a paper reporting the Ni- or Pd-catalyzed reaction of arylzinc derivatives with aryl halides.[5] The Ni- or Pd-catalyzed cross coupling reaction[6] represents one of the most general and satisfactory routes to unsymmetrical biaryls.

The currently available data, such as those summarized in Table I, indicate the following. First, Ni-phosphine and Pd-phosphine complexes may be used interchangeably in many cases (Entries 1 and 2). In cases where sterically hindered aryl reagents, such as mesitylzinc chloride, are used, Ni catalysts tend to lead to higher product yields than the corresponding Pd catalysts (Entries 11 and 12). The scope with respect to the halogen leaving group of aryl halides is broader with Ni catalysts than with Pd catalysts. As a general rule, Ni-catalyzed aryl-aryl cross coupling proceeds smoothly with both aryl iodides and aryl bromides, whereas aryl bromides must be activated by an electron-withdrawing group in Pd-catalyzed cross coupling. Palladium-catalyzed cross coupling, however, is considerably more chemoselective than Ni-catalyzed cross coupling. Thus, for example, the nitro group appears to be totally incompatible with Ni-phosphine catalysts (Entry 9), and the presence of an alkynyl group tends to lower significantly the yield of product.

Second, the choice of metal counterion is of critical importance. Except in some special cases, alkali metals such as Li, Na, and K are unsatisfactory, partly because arylmetals which contain these metals readily participate in halogen-metal exchange leading to cross-homo scrambling,[5] and also because these organometals are among the least chemoselective. Zinc appears to be among the most satisfactory metals from the standpoint of (a) product yield, (b) cross/homo ratio, (c) chemoselectivity, and (d) ease of preparation of arylmetal reagents, although Mg has also been used successfully in many cases.[7] Results shown in Entries 11 and 13 indicate that the mesitylmagnesium reagent, generated in situ by treatment of mesityllithium with magnesium bromide, appears to be considerably inferior to mesitylzinc chloride generated in a similar manner. On the other hand, mesitylmagnesium bromide generated by treatment of mesityl bromide with Mg is as effective as mesitylzinc chloride (Entry 14). However, the yield of mesitylmagnesium bromide itself is in the range 50-60% and is substantially lower than that of mesitylzinc chloride (ca. 90%). Recent results obtained with arylboron derivatives[8] appear highly promising, although the preparation of arylboronic acids is, at present, more elaborate than that of Grignard reagents or in situ generation of arylzinc reagents. Various other metals, such as Cd,[9] Hg,[10] Al,[5] Sn,[10] Zr,[9] and Cu,[10] have been shown to participate in aryl-aryl cross coupling. Their advantages over Zn or Mg are, however, largely unknown.

Third, the procedure described above has been applied to the preparation of various biaryls containing hetero-aromatic rings (Entries 15-17, 19, 20). Although the number of papers reporting the use of the Ni- or Pd-catalyzed procedure for aryl-aryl coupling is still small, the synthesis of steganone by Raphael and his co-workers[11] is demonstrative of its synthetic potential.

1. Department of Chemistry, Purdue University, W. Lafayette, IN 47907.

2. (a) Watson, S. C.; Eastham, J. F. *J. Organometal. Chem.* **1967**, *9*, 165; (b) Lipton, M. F.; Sorensen, C. M.; Sadler, A. C.; Shapiro, R. H. *J. Organometal. Chem.* **1980**, *186*, 155.

3. Coulson, D. R. *Inorg. Synth.* **1972**, *13*, 121.

4. Robinson, G. E.; Thomas, C. B.; Vernon, J. M. *J. Chem. Soc. (B)* **1971**, 1273.

5. Negishi, E.; King, A. O.; Okukado, N. *J. Org. Chem.* **1977**, *42*, 1821.

6. Negishi, E. *Acc. Chem. Res.* **1982**, *15*, 340.

7. (a) Tamao, K.; Sumitani, K.; Kiso, S.; Zembayashi, M.; Fujioka, A.; Kodama, S.; Nakajima, I.; Minato, A.; Kumada, M. *Bull. Chem. Soc. Jpn.* **1976**, *49*, 1958; (b) Kumada, M. *Pure Appl. Chem.* **1980**, *52*, 669.

8. Miller, R. B.; Dugar, S. *Organometallics* **1984**, *3*, 1261.

9. Unpublished results obtained in our laboratories.

10. Beletskaya, I. P. *J. Organometal. Chem.* **1983**, *250*, 551.

11. Larson, E. R.; Raphael, R. A. *Tetrahedron Lett.* **1979**, 5041.

TABLE I

PREPARATION OF BIARYLS BY THE Ni- OR Pd-CATALYZED REACTION OF ARYLMETALS WITH ARYL HALIDES[a]

Entry	Ar^1M[b]	ArX	Catalyst[c]	Amount (%)	Yield[d] (%)
1	PhZnCl	p-Iodoanisole	Ni	5	85
2	PhZnCl	p-Iodoanisole	Pd[e]	5	87
3	PhMgBr	p-Iodoanisole	Pd[e]	5	71
4	$PhAl(i\text{-}Bu)_2$	p-Iodoanisole	Pd[e]	5	72
5	PhZnCl	$p\text{-}BrC_6H_4CN$	Ni	5	90
6	PhZnCl	$p\text{-}BrC_6H_4COOMe$	Ni	5	70
7	PhZnCl	$p\text{-}IC_6H_4NO_2$	Pd[e]	5	90 (74)
8	o-TolZnCl	$p\text{-}BrC_6H_4NO_2$	Pd[e]	1	90 (70)
9	o-TolZnCl	$p\text{-}BrC_6H_4NO_2$	Ni	1	0
10	m-TolZnCl	$m\text{-}IC_6H_4CH_3$	Ni	5	95
11	MesZnCl[f]	$o\text{-}IC_6H_4CH_3$	Ni	5	93
12	MesZnCl	$o\text{-}IC_6H_4CH_3$	Pd	5	88
13	MesMgBr[g]	$o\text{-}IC_6H_4CH_3$	Ni	5	38
14	MesMgBr	$o\text{-}IC_6H_4CH_3$	Ni	5	92
15	PhZnCl	2-Furyl iodide	Pd	5	91
16	2-FurylZnCl	PhI	Pd	5	94 (89)
17	3-FurylZnCl	PhI	Pd	5	89 (85)
18	PhZnCl	2-Furyl bromide	Pd	5	0
19	2-ThienylZnCl	PhI	Pd	5	81 (75)
20	PhZnCl	2-Pyridyl bromide	Pd	5	99 (89)
21	PhZnCl	3-Pyridyl bromide	Pd	5	0

[a]The reactions are carried out in THF at room temperature. The time required for completion is usually less than several hours. [b]Unless otherwise mentioned, arylzinc chlorides and arylalanes are prepared via in situ transmetalation of aryllithiums, while arylmagnesium halides are prepared by treating aryl halides with Mg. The molar ratio of an aryl metal to an aryl halide is 1-1.5. [c]Ni = $Ni(PPh_3)_4$ prepared in situ by the reaction of $Ni(acac)_2$, PPh_3, and $(i\text{-}Bu)_2AlH$ (1:4:1). Unless otherwise indicated, Pd = $Pd(PPh_3)_4$. [d]By GLC. The numbers in parentheses are isolated yields. [e]The Pd catalyst is prepared by treating $Cl_2Pd(PPh_3)_2$ with $(i\text{-}Bu)_2AlH$ (2 equiv). [f]Mes = mesityl. [g]Generated in situ by treating MesLi with $MgBr_2$ generated from 1,2-dibromoethane and Mg.

Appendix

Chemical Abstracts Nomenclature (Collective Index Number); (Registry Number)

4-Methyl-4'-nitrobiphenyl: Biphenyl, 2-methyl-4'-nitro- (8); 1,1'-Biphenyl, 2-methyl-4'-nitro- (9); (33350-73-1)

o-Tolylzinc chloride: Zinc, chloro(2-methylphenyl)- (11); (84109-17-1)

o-Iodotoluene: Toluene, o-iodo- (8); Benzene, 1-iodo-2-methyl- (9); (615-37-2)

Tetrakis(triphenylphosphine)palladium: Palladium, tetrakis(triphenylphosphine)- (8); Palladium, tetrakis(triphenylphosphine)-, (T-4)- (9); (14221-01-3)

VINYL RADICAL CYCLIZATION VIA ADDITION OF TIN RADICALS TO TRIPLE BONDS

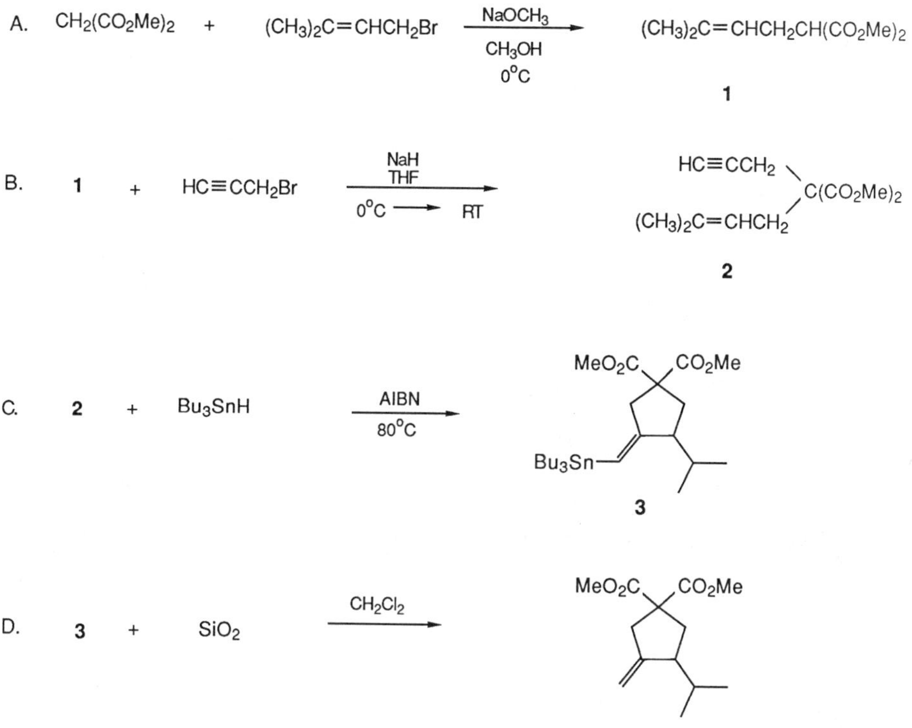
A. CH2(CO2Me)2 + (CH3)2C=CHCH2Br
NaOCH3
CH3OH
0°C
(CH3)2C=CHCH2CH(CO2Me)2
1
B. 1 + HC≡CCH2Br
NaH
THF
0°C → RT
HC≡CCH2
C(CO2Me)2
(CH3)2C=CHCH2
2
C. 2 + Bu3SnH
AIBN
80°C
MeO2C
CO2Me
Bu3Sn
3
D. 3 + SiO2
CH2Cl2
MeO2C
CO2Me

4

Submitted by Robert Mook, Jr. and Philip Michael Sher.[1]
Checked by Anthony G. Schaefer and Leo A. Paquette.

1. Procedure

A. (3-Methyl-2-butenyl)propanedioic acid, dimethyl ester (**1**). To a 500-mL, flame-dried, three-necked, round-bottomed flask under an argon atmosphere and fitted with a thermometer, pressure-equalizing addition funnel and a magnetic stirring bar, is added 250 mL of methanol (Note 1). The flask is immersed in an ice bath, and 6.7 g (0.29 mol) of sodium is added cautiously (Note 2). After the sodium has dissolved, the ice bath is removed, 36.9 g (0.28 mol) of dimethyl malonate (Note 1) is added at room temperature, and the solution is stirred for 0.5 hr. The reaction mixture is cooled to 0°C, and 45.8 g (0.31 mol) of 3,3-dimethylallyl bromide (Note 1) is added dropwise while the temperature is maintained near 5°C (Note 3). After 1 hr (Note 4), the reaction mixture is transferred to a 1-L, one-necked, round-bottomed flask with the aid of a small amount of methanol and concentrated with a rotary evaporator. The white residue is taken up in 400 mL of ether and 300 mL of a saturated salt/saturated sodium bicarbonate (1:1) solution and is transferred to a separatory funnel. The ether layer is separated, and the aqueous layer is extracted with ether (1 x 200 mL). The ether layers are combined, dried over magnesium sulfate, filtered, and concentrated with a rotary evaporator. Distillation of the residue through a 6-inch Vigreux column (after a small fore-run is collected) yields 43.9-44.5 g (78-79%) of **1**, bp 60-63°C (0.15 mm) (Note 5).

B. (3-Methyl-2-butenyl)(2-propynyl)propanedioic acid, dimethyl ester (**2**). To a 1-L, flame-dried, three-necked, round-bottomed flask, equipped with a magnetic stirring bar (Note 6) and under an argon atmosphere, is added 9.0 g (0.19 mol) of sodium hydride dispersion (Note 7). The sodium hydride is washed with petroleum ether (4 x 30 mL), removing the petroleum ether by pipette after the sodium hydride has settled. The flask is then fitted with a thermometer and an oven-dried pressure-equalizing addition funnel and charged with 500 mL of tetrahydrofuran (Note 7). The heterogenous mixture is cooled with an ice bath, and 36.4 g (0.18 mol) of the monoalkyl diester (**1**) is added dropwise at the rate of 1 drop/2-3 sec (Note 2). The cooling bath is removed when the addition is complete, and the solution is stirred until no more gas evolves (approximately 1 hr). The reaction mixture is recooled to 0°C, and 22 mL of propargyl bromide solution (0.20 mol) (Note 7) is added dropwise while the temperature is maintained at 0-10°C. Sodium bromide begins to precipitate within 20 min. The ice bath is removed, and the reaction mixture is stirred overnight (Note 8). After careful addition of 50 mL of water (Note 9) and removal of the stirring bar, the solution is transferred to a 1-L, one-necked, round-bottomed flask and concentrated with a rotary evaporator. The residue is taken up in 500 mL of ether and washed with water (3 x 300 mL) and saturated salt solution (1 x 100 mL). The aqueous layers are combined, saturated with salt, and extracted with ether (2 x 150 mL). The ether layers are combined, dried over magnesium sulfate, filtered, and concentrated with a rotary evaporator. The residue is distilled through a short path distillation apparatus at 80°C (0.25 mm) to yield 34.0-34.2 g (79-80%) of 2 (Note 10).

C. *(Z)-3-Tributylstannylmethylene-4-isopropyl-1,1-cyclopentanedicarboxylic acid, dimethyl ester* (**3**). A flame-dried, 100-mL, round-bottomed flask equipped with a magnetic stirring bar is charged with argon, and 23.8 g (0.100 mol) of dialkylmalonate diester (**2**), 30.2 g (< 0.104 mol) of tributyltin hydride (Note 11), and 40 mg (0.25 mmol) of azobisisobutyronitrile (AIBN) (Note 12) are added neat (Notes 13, 14). The entire assembly is lowered into an oil bath maintained at 75-85°C, and the mixture is stirred. After an induction period of less than 30 min, an exothermic reaction occurs which produces a small amount of gas and a rise in the temperature of the oil bath (as much as 20°C for a small bath). After this point TLC shows that the reaction is essentially complete (Notes 15, 16, 17). Unpurified 3 thus obtained is suitable for protodestannylation.

D. 3-Methylene-4-isopropyl-1,1-cyclopentanedicarboxylic acid, dimethyl ester (**4**). Crude vinylstannane (**3**) is transferred to a 2-L Erlenmeyer flask which contains 1 L of dichloromethane, 350 g of silica (Note 18), and a large (7 cm x 3 cm egg-shaped) stirring bar. The flask is stoppered, and the mixture is stirred for 24-48 hr (Note 19). The mixture is divided into three portions. Each portion is filtered under reduced pressure with a 600-mL glass frit, and the silica is washed with ethyl acetate (8 x 200 mL) to extract all of the desired product (Note 20). The solution is filtered through Celite and the solvent is removed with a rotary evaporator. Distillation through a short path distillation apparatus (with no forerun) gives 19.8-20.5 g of **4**, (83-85% overall from 2), bp 80°-85°C (0.2 mm) (Note 21).

2. Notes

1. The use of less solvent can result in gel formation. Methanol was freshly distilled from calcium hydride or magnesium metal. Dimethyl malonate (Aldrich Chemical Company, Inc.) was distilled before use. 3,3-Dimethyallyl bromide may be purchased from Aldrich Chemical Company, Inc. or Wiley Organics.

2. The apparatus should be vented. Hydrogen gas formation causes vigorous bubbling.

3. At higher temperatures more dialkylation occurs.

4. The reaction can be monitored by TLC eluting with 10% ethyl acetate/petroleum ether.

5. The physical properties are as follows: IR (neat) cm^{-1}: 2980, 1730-1760, 1435, 1040; ^{1}H 200 MHz NMR ($CDCl_3$) δ: 1.62 (s, 3 H), 1.65 (s, 3 H), 2.56 (t, 2 H, J = 7), 3.34 (t, 1 H, J = 7), 3.70 (s, 6 H), 5.06 (bt, 1 H, J = 7).

6. Unless a sufficiently large stirring bar is used, the precipitate may be impossible to stir. An overhead mechanical stirrer can be substituted.

7. Sodium hydride, 50% oil dispersion, was purchased from Alfa Products, Morton/Thiokol Inc. Tetrahydrofuran was freshly distilled from sodium/benzophenone. Propargyl bromide (80% in toluene) was purchased from Aldrich Chemical Company, Inc. and used directly.

8. The reaction is generally complete sooner, but because its progress is difficult to monitor by TLC (since the starting material and product have similar R_fs) the submitters routinely allow more than enough time.

9. Trace amounts of sodium hydride may still be present. If the water is added too quickly a vigorous reaction results.

10. The checkers observed a boiling point of 99-100°C (0.35 mm) for this material. The physical properties are as follows: IR (neat) cm^{1}: 3290, 2960, 1740, 1440; ^{1}H 300 MHz NMR ($CDCl_3$) δ: 1.61 (s, 3 H), 1.65 (s, 3 H), 1.97 (t, 1 H, J = 3), 2.7 (m, 4 H), 3.69 (s, 6 H), 4.9 (t, 1 H, J = 7).

11. Tributyltin hydride, 97%, was purchased from Aldrich Chemical Company, Inc. and used directly. A minimal excess of this reagent is desired in order to ensure clean distillation of **4**. If the product is to be purified by chromatography, a 10% excess of tributyltin hydride can be used.

12. Azobisisobutyronitrile (AIBN) was purchased from Aldrich Chemical Company, Inc. and recrystallized from chloroform. To exclude the possibility of a violent reaction, no more than twice this amount should be used on this scale.

13. These reactions can also be run in benzene, which is preferable in cases where transfer of hydrogen to the vinyl radical is competitive with cyclization.

14. The submitters routinely interposed a Vigreux column between the reaction flask and the argon line to protect the argon line in the event of bumping.

15. The reaction mixture is routinely stirred hot for an additional 30 min to ensure complete conversion.

16. Vinylstannane (**3**) may protodestannylate on TLC, which could be confusing when monitoring reaction C since dialkylmalonate diester (**2**) and protodestannylated product (**4**) have similar R_fs. The submitters found that TLC plates stored in the open air (as opposed to desiccator-stored plates) caused negligible protodestannylation.

17. Although vinylstannane **3** protodestannylates on silica, it may be isolated by flash chromatography in greater than 90% yield: IR (neat) cm^{-1}: 2960, 2930, 2880, 2860, 1740, 1615, 1465, 1435; ^1H 200 MHz NMR ($CDCl_3$) δ: 3.71 (s, 3 H), 3.73 (s, 3 H), 5.62 (bs, 1 H) with satellites.

18. The submitters employed Silica Woelm 32-63 purchased from Universal Scientific, Inc.; the checkers used Merck silica gel 60 (40-60 μm in size). Before use the silica was oven dried for several days at 160°C. Undried silica may be used, but larger quantities and/or longer reaction times are necessary.

19. TLC plates stored in the open air should be used (see Note 16) to monitor reaction D accurately. Approximately 4% of noncyclized hydrostannylation product is produced in reaction C. This material forms co-spots with **3** (2.5% ethyl acetate in petroleum ether), but protodestannylates much more slowly. Therefore, by TLC reaction D may appear not to go to completion.

20. A larger glass frit would obviate division of the mixture. In any case, the silica should be washed until by TLC the filtrate no longer contains **4**.

21. The checkers observed a boiling point of 98-105°C (0.35 mm) for this product: IR (neat) cm^{-1}: 3080, 2960, 2875, 1735, 1655, 1435, 1385, 1365, 890. MS (CI) M + 1 = 241; ^{1}H 300 MHz NMR ($CDCl_3$) δ: 0.75 (d, 3 H, J = 7), 0.85 (d, 3 H, J = 7), 1.80 (m, 2 H), 2.40 (m, 2 H), 2.84 (m, 2 H), 3.64 (s, 3 H), 3.65 (s, 3 H), 4.72 (bs, 1 H), 4.90 (bs, 1 H).

3. Discussion

Free radical reactions are proving to be synthetically useful alternatives for producing carbon-carbon bonds.[2,3] Recently, Stork has shown that vinyl radicals are valuable in ring forming reactions since they place a double bond in a predictable position.[3] Their compatibility with many unprotected functional groups and their ability to form quaternary centers are additional features which make vinyl radical cyclization an attractive synthetic method.

Previously, vinyl radicals for cyclization reactions were produced by the reduction of vinyl halides with tributyltin hydride. In the present procedure, vinyl radicals are produced by the addition of tin radicals to triple bonds.[4] These vinyl radicals undergo cyclization in an analogous fashion to those generated from vinyl halides. However, this approach provides vinylstannanes stereoselectively; these vinylstannanes may be utilized in a wide range of synthetic transformations.

Vinylstannanes[5] are versatile synthetic intermediates. They serve as a source of stereospecific vinyl anions and vinyl cuprates.[5a,d,6] In the presence of Pd(0), vinylstannanes can be acylated[7a,b] or alkylated.[7c,d] Epoxidation followed by rearrangement converts vinylstannanes into carbonyl compounds.[8] Treatment with halogens or N-halosuccinimides yields vinyl halides.[5a-d,6c,9] Deuterium labeled olefins result from deuterolysis of vinyl stannanes with AcOD or DCl.[5d] Olefins identical to those produced in vinyl halide reductions are obtained on protodestannylation with silica gel, which, unlike simple acid treatment, solves the problem of separation from tin residues.[10]

Many examples of radicals kinetically favoring addition to double bonds over triple bonds are known.[11] Yet, vinyl radical cyclization in the present procedure is initiated by the addition of a tin radical to a triple bond.[12] The apparent selectivity of the tin radical for the triple bond in the presence of a double bond is, at least in some cases, a result of reversible addition to both followed by selective cyclization of the vinyl radical.[4]

1. Department of Chemistry, Columbia University, New York, NY 10027. We greatly appreciate the guidance and support we received from Professor Gilbert Stork during the course of this work.

2. (a) Hart, D. J. *Science* **1984**, *223*, 883 and references within; (b) Barton, D. H. R.; Crich, D. *Tetrahedron Lett.* **1984**, *25*, 2787; (c) Giese, B. *Angew. Chem., Intern. Ed. Engl.* **1983**, *22*, 753 and references within; Giese, B.; Gonzalez-Gomez, J. A.; Witzel, T. *Angew. Chem., Intern. Ed. Engl.* **1984**, *23*, 69; (d) Beckwith, A. L. J; O'Shea, D. M.; Roberts, D. H. *J. Chem. Soc., Chem. Commun.* **1983**, 1445; (e) Keck, G. E.; Enholm, E. J. *Tetrahedron Lett.* **1985**, *26*, 3311 and references within; (f) Minisci, F.; Citterio, A.; Giordano, C. *Acc. Chem. Res.* **1983**, *16*, 27 and references within; Minisci, F. *Top. Curr. Chem.* **1976**, *62*, 1; Minisci, F.; Citterio, A. In "Advances in Free-Radical Chemistry", Williams, G. H., Ed.; Heyden: London, 1980; Vol. 6, p. 65; (g) Angoh, A. G.; Clive, D. L. J. *J. Chem. Soc., Chem. Comm.* **1985**, 980, and references within; (h) Curran, D. P.; Rakiewicz, D. M. *J. Am. Chem. Soc.* **1985**, *107*, 1448; (i) Wilcox, C. S.; Thomasco, L. M. *J. Org. Chem.* **1985**, *50*, 546.

3. (a) Stork, G.; Kahn, M. *J. Am. Chem. Soc.* **1985**, *107*, 500; (b) Stork, G. In "Selectivity - A Goal for Synthetic Efficiency"; Bartmann, W.; Trost, B. M., Eds.; Verlag Chemie: Weinheim, 1984, p. 281; (c) Stork, G. In "Current Trends in Organic Synthesis"; Nozaki, H., Ed.; Pergamon Press: New York, 1983, p. 359; (d) Stork, G.; Sher, P. M. *J. Am. Chem. Soc.* **1983**, *105*, 6765; (e) Stork, G.; Mook, R., Jr. *J. Am. Chem. Soc.* **1983**, *105*, 3720; (f) Stork, G.; Baine, N. H. *J. Am. Chem. Soc.* **1982** *104*, 2321; (g) Stork, G.; Willard, P. G. *J. Am. Chem. Soc.* **1977**, *99*, 7067.

4. (a) Stork, G.; Mook, R., in preparation, 1985; (b) Mook, R., Ph.D. Dissertation, Columbia University, N.Y., 1985.

5. (a) Negishi, E.-i. "Organometallics in Organic Synthesis"; John Wiley-Interscience Books: New York, 1980; Vol. 1, pp. 394-454; (b) Poller, R. C. "The Chemistry of Organotin Compounds"; Academic Press: New York, 1970; Chapters 2, 3, 7; (c) Neumann, W. P. "The Organic Chemistry of Tin"; John Wiley and Sons: New York, 1970; Chapters 3, 4, 8, 9; (d) Pereyre, M.; Quintard, J.-P. *Pure Appl. Chem.* **1981**, *53*, 2401 and references within; (e) Baldwin, J. E.; Kelly, D. R.; Ziegler, C. B. *J. Chem. Soc., Chem. Commun.* **1984**, 133; (f) Seitz, D. E.; Lee, S.-H. *Tetrahedron Lett.* **1981**, *22*, 4909; (g) Piers, E.; Chong, J. M. *J. Chem. Soc., Chem. Commun.* **1983**, 934; (h) Westmijze, H.; Ruitenberg, K.; Meijer, J.; Vermeer, P. *Tetrahedron Lett.* **1982**, *23*, 2797; (i) Masure, D.; Coutrot, Ph.; Normant, J. F. *J. Organomet. Chem.* **1982**, *226*, C 55; (j) Hibino, J.-I.; Matsubara, S.; Morizawa, Y.; Oshima, K.; Nozaki, H. *Tetrahedron Lett.* **1984**, *25*, 2151; (k) Taylor, R. T.; Degenhardt, C. R.; Melega, W. P.; Paquette, L. A. *Tetrahedron Lett.* **1977**, 159; (l) Shibasaki, M.; Torisawa, Y.; Ikegami, S. *Tetrahedron Lett.* **1982**, *23*, 4607.

6. (a) Seyferth, D.; Weiner, M. A. *J. Am. Chem. Soc.* **1961**, *83*, 3583 and references within; (b) Corey, E. J.; Wollenberg, R. H. *J. Am. Chem. Soc.* **1974**, *96*, 5581 and *J. Org. Chem.* **1975**, *40*, 2265; (c) Ensley, H. E.; Buescher, R. R.; Lee, K. *J. Org. Chem.* **1982**, *47*, 404 and references within.

7. (a) Labadie, J. W.; Stille, J. K. *Tetrahedron Lett.* **1983**, *24*, 4283 and references within; (b) For acylation without the use of Pd(0) see references 5d; (c) Kosugi, M.; Hagiwara, I.; Migita, T. *Chem. Lett.* **1983**, [illegible] 839; (d) Scott, W. J.; Crisp, G. T.; Stille, J. K. *J. Am. Chem. Soc.* **1984**, *106*, 4630.

8. Nishida, A.; Shibasaki, M.; Ikegami, S. *Tetrahedron Lett.* **1981**, *22*, 4819.

9. Jung, M. E.; Light, L. A. *Tetrahedron Lett.* **1982**, *23*, 3851 and references within.

10. Leibner, J. E.; Jacobus, J. *J. Org. Chem.* **1979**, *44*, 449 and references within.

11. (a) Kharasch, M. S.; Jerome, J. J.; Urry, W. H. *J. Org. Chem.* **1950**, *15*, 966; (b) Gazith, M.; Szwarc, M. *J. Am. Chem. Soc.* **1957**, *79*, 3339; (c) Abell, P. I. In "Comprehensive Chemical Kinetics"; Bamford, C. H.; Tripper, C. F. H., Eds.; Elsevier: Amsterdam, 1976; Vol. 18, p. 211; (d) Giese, B.; Lachhein, S. *Angew Chem., Intern. Ed. Engl.* **1982**, *21*, 768.

12. The sterically less biased case of allylpropargyl malonate gives greater than 75% of analogous product.

Appendix

Chemical Abstracts Nomenclature (Collective Index Number); (Registry Number)

(3-Methyl-2-butenyl)propanedioic acid, dimethyl ester: Propanedioic acid, (3-methyl-2-butenyl)-, dimethyl ester (9); (43219-18-7)

Methanol (8,9); (67-56-1)

Sodium (8,9); (7440-23-5)

Dimethyl malonate: Malonic acid, dimethyl ester (8); Propanedioic acid, dimethyl ester (9); (108-59-8)

3,3-Dimethylallyl bromide: 2-Butene, 1-bromo-3-methyl- (8,9); (870-63-3)

Sodium hydride (8,9); (7646-69-7)

Propargyl bromide: Propyne, 3-bromo- (8); 1-Propyne, 3-bromo- (9); (106-96-7)

Tributyltin hydride: Stannane, tributyl- (8,9); (688-73-3)

Azobisisobutyronitrile: Propionitrile, 2,2'-azobis[2-methyl- (8);

Propanenitrile, 2,2'-azobis[2-methyl- (9); (78-67-1)

CYCLOPENTANONES FROM CARBOXYLIC ACIDS VIA INTRAMOLECULAR ACYLATION OF ALKYLSILANES: 2-METHYL-2-VINYLCYCLOPENTANONE

(Cyclopentanone, 2-ethenyl-2-methyl-)

A. Me_3Si–$CH_2CH_2CH_2$–Cl $\xrightarrow{NaI}$ Me_3Si–$CH_2CH_2CH_2$–I (**1**)

B. CO_2Me $\xrightarrow{1)\ LDA;\ 2)\ \mathbf{1}}$ CO_2Me, $SiMe_3$ (**2**)

CO_2Me, $SiMe_3$ $\xrightarrow{1)\ KOH;\ 2)\ H^+}$ CO_2H, $SiMe_3$ (**3**)

CO_2H, $SiMe_3$ $\xrightarrow{1)\ (COCl)_2;\ 2)\ AlCl_3}$ O

Submitted by Isao Kuwajima and Hirokazu Urabe.[1]

Checked by Robert J. Ross and Leo A. Paquette.

1. Procedure

Caution! The following reactions should be performed in an efficient hood.

A. 1-Iodo-3-trimethylsilylpropane. In a dry, 100-mL, two-necked, round-bottomed flask equipped with a magnetic stirring bar, reflux condenser, and rubber septum is placed 15.0 g (0.10 mol) of sodium iodide. A nitrogen inlet tube is connected to the top of the reflux condenser and all the apparatus is kept under nitrogen. To this vessel are added 50 mL of acetone and 11.5 mL

(10 g, 0.066 mol) of 1-chloro-3-trimethylsilylpropane (Note 1) with a hypodermic syringe through the septum; the resulting white suspension is stirred under reflux for 24 hr. The condenser is replaced with a Claisen head, and the bulk of the solvent is removed under ordinary pressure (Note 2) to give a white slurry. To this is added 60 mL of hexane and the inorganic salts are filtered off by suction. The filter cake is washed with five 10-mL portions of hexane. The hexane is distilled off from the combined organic portions at atmospheric pressure. The residual oil is transferred to a 50-mL, round-bottomed flask fitted with a stirring bar and a Claisen head, and distilled under reduced pressure to afford 1-iodo-3-trimethylsilylpropane 1 (11.5-13.1 g, 72-81%), bp 84-86°C (25 mm), as a clear liquid (Notes 3 and 4).

B. 2-Methyl-2-vinylcyclopentanone. A 300-mL, two-necked, round-bottomed flask fitted with a magnetic stirring bar, nitrogen inlet tube, and rubber septum is kept under dry nitrogen. To this flask are introduced 8.3 mL (5.99 g, 0.0591 mol) of diisopropylamine and 120 mL of tetrahydrofuran (Note 5) with a syringe through the septum. The flask is cooled in a dry ice-hexane bath. To the solution is slowly added 38.7 mL of butyllithium (1.53 M in hexane, 0.0592 mol) with a syringe and the mixture is kept at 0°C (in an ice bath) for 10 min. The resulting solution of lithium diisopropylamide is again cooled in a dry ice-hexane bath (-78°C), and 11.2 mL (11.57 g, 0.0645 mol) of hexamethylphosphoric triamide (HMPA) (Note 6) is added. After stirring for 30 min, 6.47 mL (6.15 g, 0.0539 mol) of methyl tiglate is injected drop by drop at -78°C. After the solution is stirred for an additional 20 min, 13.0 g (0.0537 mol) of 1-iodo-3-trimethylsilylpropane 1 is added with a syringe and the dry ice-hexane bath is replaced with an ice bath. The solution is stirred at about 0°C for 1 hr and then poured onto 150 mL of ice-cooled 3 N hydrochloric acid covered with 150 mL of hexane. The organic layer is separated and the aqueous layer is extracted with two 80-mL portions of hexane. The combined organic layers are washed successively with 50 mL of 3 N hydrochloric acid and 50 mL of saturated sodium bicarbonate solution, and dried over anhydrous magnesium sulfate. The solvent is removed on a rotary evaporator to afford the crude ester **2** (ca. 13 g) (Note 7) which is sufficiently pure for the next operation.

In a two-necked, round-bottomed flask equipped with a magnetic stirring bar, rubber septum, and reflux condenser, the top of which is fitted with a nitrogen inlet tube, is placed 7.3 g (ca. 0.13 mol) of 85% pure potassium hydroxide. To the flask are added 4 mL of water and an ethanol solution (60 mL) of the crude ester **2** (ca. 13 g) with a syringe; the mixture is refluxed for 1 hr. The reflux condenser is replaced with a Claisen head and the bulk of the solvent is distilled off over 30 min (Note 8). The residue is cooled in an ice bath and 80 mL of 6 N hydrochloric acid is cautiously introduced. The mixture is extracted with 150 mL of hexane and the organic layer is separated. To the aqueous layer is added 40 mL of concentrated hydrochloric acid and the solution is again extracted with two 100-mL portions of hexane. The combined organic extracts are dried over anhydrous magnesium sulfate and concentrated under reduced pressure to afford the crude acid 3 (10.6-12.0 g) as a dark-colored oil (Note 9).

A 200-mL, two-necked, round-bottomed flask fitted with a magnetic stirring bar, reflux condenser, and rubber septum is flushed with nitrogen. To this flask is introduced a dry benzene solution (50 mL) of crude acid 3 through the septum. Then 8.9 mL (12.9 g, 0.101 mol) of oxalyl chloride is slowly added with stirring at room temperature. After the evolution of gas ceases, the solution is further heated in an oil bath maintained at 70°C for 30 min. The solvent, together with excess oxalyl chloride, is removed at room temperature first on a rotary evaporator and finally with a vacuum pump to leave the crude acid chloride of **3** as a dark-colored oil (Note 10).

In a 300-mL, three-necked, round-bottomed flask fitted with a nitrogen inlet tube, dropping funnel, and rubber septum are placed 7.45 g (0.0559 mol) of powdered aluminum chloride and a magnetic stirring bar. After 100 mL of dichloromethane (Note 11) is introduced, the crude acid chloride, dissolved in 50 mL of dichloromethane, is added via the dropping funnel to the stirred suspension of aluminum chloride at 0°C over 5 min, whereupon most of the aluminum chloride dissolves. After further stirring at 0°C for 15 min and at room temperature for 15 min, the flask is recooled in an ice bath and 100 mL of 3 N hydrochloric acid is cautiously added through the dropping funnel. The organic phase is separated and the aqueous layer is extracted three times with 50-mL portions of dichloromethane. The combined organic layers are successively washed with 30 mL of 3 N hydrochloric acid and 50 mL of saturated sodium bicarbonate solution, and dried over anhydrous magnesium sulfate. The solvent is removed under ordinary pressure and the residue is distilled under reduced pressure to give 2-methyl-2-vinylcyclopentanone as a clear liquid (3.74-5.6 g, 56-84% yield based on the methyl tiglate), bp 104-124°C (110 mm) (Note 12), ca. 95% pure by GLC (Note 13).

2. Notes

1. 1-Chloro-3-trimethylsilylpropane, obtained from Petrarch Systems, Inc., was used as received.

2. About 40 mL of distillate is collected.

3. 1-Iodo-3-trimethylsilylpropane has the following spectral properties: ^{1}H NMR (CCl_4, 3% benzene (δ 7.24) as an internal standard) δ: -0.03 (s, 9 H, $(CH_3)_3Si$), 0.34-0.74 (m, 2 H, CH_2Si), 1.47-2.04 (m, 2 H, CH_2), 3.07 (t, 2 H, J = 7, CH_2I); IR (neat) cm^{-1}: 2950, 1250, 860, 830.

4. 1-Iodo-3-trimethylsilylpropane can also be prepared by other methods: See reference 2.

5. Tetrahydrofuran was used after distillation from sodium benzophenone ketyl under nitrogen.

6. It was found by the checkers that N,N'-dimethylpropyleneurea (DMPU),[3] supplied by Aldrich Chemical Company, Inc. or Fluka, can be substituted for HMPA with no change in either procedure or yield.

7. Alkylation of methyl tiglate was carried out according to a reported procedure.[4]

8. About 50 mL of distillate was collected.

9. Crude 3 exhibited the following spectral properties, which are virtually identical with those of an analytically pure sample: ^{1}H NMR (CCl_4, 3% benzene (δ 7.24) as an internal standard) δ: 0.16 (s, 9 H, $(CH_3)_3Si$), 0.44-0.77 (m, 2 H, CH_2Si), 1.1-2.0 (m, 4 H, CH_2CH_2), 1.41 (s, 3 H, CH_3), 4.91-5.31 (m, 2 H, $C{=}CH_2$), 6.04 (d of d, 1 H, J = 10 and 18, $CH{=}CH_2$), 11.34 (s, 1 H, CO_2H); IR (neat) cm^{-1}: 2950, 1700, 1400, 1250, 1180, 920, 840, 740, 680.

10. Conversion of the carboxylic acid to the acid chloride was based on a reported method.[5]

11. Dichloromethane was distilled from phosphorus pentoxide.

12. 2-Methyl-2-vinylcyclopentanone showed the following spectral properties: ^{1}H NMR (CCl_4) δ: 0.73 (s, 3 H, CH_3), 1.6-2.3 (m, 6 H, $(CH_2)_3$), 4.67-5.07 (m, 2 H, $C{=}CH_2$), 5.62 (d of d, 1 H, J = 8 and 18, $CH{=}CH_2$); IR (neat) cm^{-1}: 3070, 2950, 1730, 1640, 1450, 1400, 1150, 1060, 1040, 1000, 920.

13. Vapor phase chromatography was performed on an OV 101, fused silica, 20-m capillary column.

3. Discussion

Cyclopentanones are widely found in natural products and are also useful intermediates in organic synthesis. Thus a facile construction of cyclopentanones from easily available acyclic precursors is particularly desirable. This method of preparation is based on an intramolecular acylation of 5-trimethylsilylalkanoyl chlorides previously reported by us.[2] The starting materials are generally prepared by alkylation of carboxylic acids with 3-trimethylsilylalkyl halides followed by their conversion to the corresponding acid chlorides. The cyclization of the acid chlorides proceeds cleanly with aluminum chloride. An acyl cation generated from the acid chloride and aluminum chloride is trapped with the alkyl-silicon bond in the same molecule to yield a cyclopentanone selectively (eq. 1).

Other results are collected in the Table,[2] which shows that cyclopentanones having a variety of substituents can be prepared according to this procedure with substantial advantage over other methods, e.g. alkylation of 2-alkoxycarbonylcyclopentanone.

TABLE

CLOPENTANONE SYNTHESIS BY INTRAMOLECULAR ACYLATION OF 5-TRIMETHYLSILYLALKANOYL CHLORIDES[a]

CO_2H / $SiMe_3$ (1) $\xrightarrow{(COCl)_2}$ COCl / $SiMe_3$ (2) $\xrightarrow{AlCl_3}$ O (3)

Acid (1)	Product (3)	Yield(%)[b]
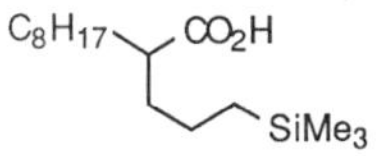 C_8H_{17}, CO_2H, $SiMe_3$	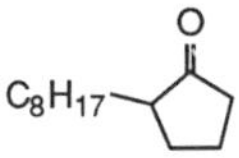 O, C_8H_{17}	92[c]
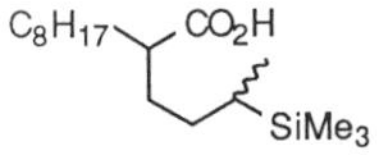 C_8H_{17}, CO_2H, $SiMe_3$	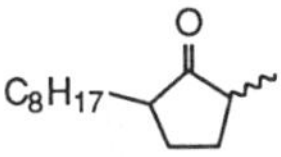 O, C_8H_{17}	83
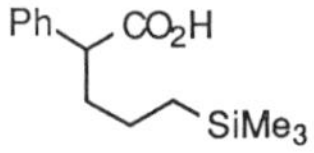 Ph, CO_2H, $SiMe_3$	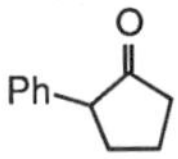 O, Ph	84
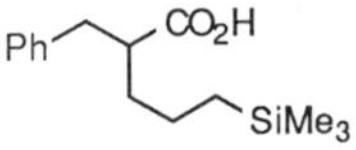 Ph, CO_2H, $SiMe_3$	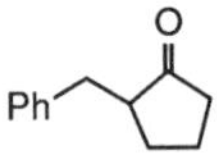 O, Ph	85
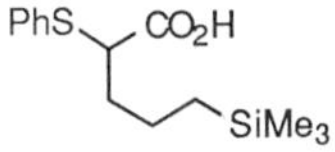 PhS, CO_2H, $SiMe_3$	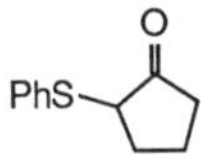 O, PhS	60[d]
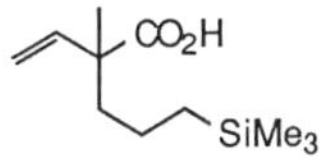 CO_2H, $SiMe_3$	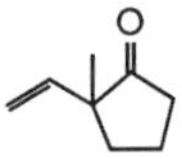 O	67
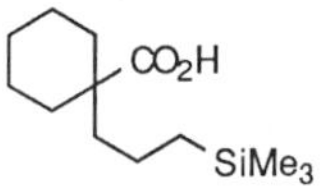 CO_2H, $SiMe_3$	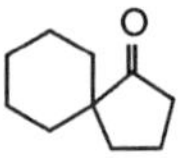 O	70
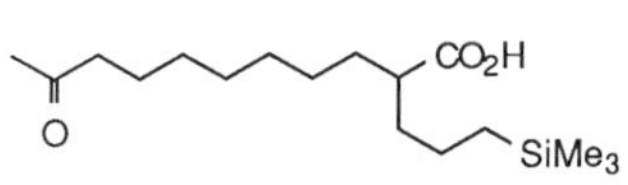 O, CO_2H, $SiMe_3$	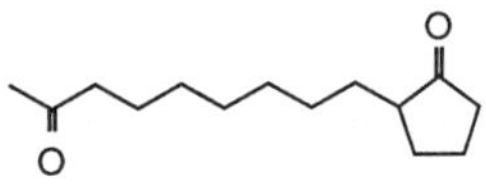 O, O	76[e]

[a] eactions are carried out on 0.2-0.3-mmol scale with the reactant ratio $1/(COCl)_2/AlCl_3$ = 1:2:1.

[b] verall yield from 1. Products are isolated by chromatography.

[c] eaction on 3.5-mmol scale; the product was isolated by Kugelrohr distillation.

[d] eactant ratio: $1/(COCl)_2/AlCl_3$ = 1:1.5:0.75.

[e] eactant ratio: $1/(COCl)_2/AlCl_3$ = 1:2:2.

1. Department of Chemistry, Tokyo Institute of Technology, Meguro, Tokyo 152, Japan.
2. Urabe, H.; Kuwajima, I. *J. Org. Chem.* **1984**, *49*, 1140.
3. Mukhopadhyay, T.; Seebach, D. *Helv. Chim. Acta* **1982**, *65*, 385.
4. Herrmann, J. L.; Kieczykowsky, G. R.; Schlessinger, R. H. *Tetrahedron Lett.* **1973**, 2433.
5. Adams, R.; Ulich, L. H. *J. Am. Chem. Soc.* **1920**, *42*, 599.

Appendix

Chemical Abstracts Nomenclature (Collective Index Number); (Registry Number)

2-Methyl-2-vinylcylopentanone: Cyclopentanone, 2-ethenyl-2-methyl- (11); (88729-76-4)

1-Iodo-3-trimethylsilylpropane: Silane, (3-iodopropyl)trimethyl- (9); (18135-48-3)

1-Chloro-3-trimethylsilylpropane: Silane, (3-chloropropyl)trimethyl- (8,9); (2344-83-4)

Hexamethylphosphoric triamide: Phosphoric triamide, hexamethyl- (8,9); (680-31-9)

N,N'-Dimethylpropyleneurea: 2(1H)-Pyrimidone, tetrahydro-1,3-dimethyl- (8,9); (7226-23-5)

Methyl tiglate: 2-Butenoic acid, 2-methyl-, methyl ester, (E)- (9); (6622-76-0)

Oxalyl chloride (8); Ethanedioyl dichloride (9); (79-37-8)

VINYLATION OF ENOLATES WITH A VINYL CATION EQUIVALENT: trans-3-METHYL-2-VINYLCYCLOHEXANONE

A. $[C_5H_5Fe(CO)_2]_2 \xrightarrow[THF]{Na(Hg)_n} C_5H_5Fe(CO)_2Na \xrightarrow[50° C]{ClCH_2CH(OEt)_2}$

$C_5H_5(CO)_2FeCH_2CH(OEt)_2 \xrightarrow[CH_2Cl_2, -78°C]{HBF_4 \cdot (C_2H_5)_2O} C_5H_5(CO)_2\overset{+}{Fe}$—($CH_2$=$CHOC_2H_5$) BF_4^-

1

B. (2-cyclohexen-1-one) $\xrightarrow[Et_2O]{Me_2CuLi}$ (lithium enolate of 3-methylcyclohexanone, OLi, Me) $\xrightarrow[THF, ether, -78°C]{1}$

(2: 2-[2-ethoxy-2-($Fe(CO)_2C_5H_5$)ethyl]-3-methylcyclohexanone, OC_2H_5, Me) $\xrightarrow[Et_2O, -78°]{HBF_4 \cdot (C_2H_5)_2O}$ (3: $\overset{+}{Fe}(CO)_2C_5H_5$ vinyl complex, Me) BF_4^-

2 3

C. 3 $\xrightarrow[reflux]{CH_3CN}$ (4: trans-3-methyl-2-vinylcyclohexanone, Me) + $C_5H_5(CO)_2Fe(CH_3CN)^+ BF_4^-$

4

Submitted by Tony C. T. Chang,[1] Myron Rosenblum,[2] and Nancy Simms.[2]

Checked by Ed Fewkes and Martin F. Semmelhack.

1. Procedure

Caution! Care should be exercised in the preparation of the sodium amalgam since the initial reaction is highly exothermic. This and all subsequent operations should be carried out in a well-ventilated hood. Chloroacetaldehyde diethyl acetal is an irritant and a mutagen. Care should be exercised in its handling.

A. Dicarbonyl(cyclopentadienyl)(ethyl vinyl ether)iron tetrafluoroborate **1.** A 500-mL, three-necked flask, with a stopcock at the bottom, is fitted with a nitrogen inlet and a mechanical stirrer with a Teflon paddle. Nitrogen is passed through the flask while it is flame-dried (Note 1) and then 70 mL of mercury is introduced. The mercury is stirred vigorously as 7.2 g (0.31 mol) of sodium metal, cut into small pieces, is slowly added, under a strong flow of nitrogen; after which the remaining neck is capped with a rubber septum. The amalgam is allowed to cool to room temperature and 100 mL of tetrahydrofuran is added (Notes 2 and 3). While the system is flushed with nitrogen, one septum is removed and 35.4 g (0.1 mol) of dicarbonyl(cyclopentadienyl)diiron $[(CO)_2CpFe]_2$ (Note 4) is added at once. Vigorous stirring is continued for 40 min. The mercury is drained through the stopcock, and the deep yellow-red solution of sodium dicarbonyl(cyclopentadienyl)ferrate, which is ready for use without further purification, is transferred (Note 2) to a 500-mL, round-bottomed flask containing a magnetic stirrer. An additional 50 mL of tetrahydrofuran is used to rinse the amalgam flask.

Chloroacetaldehyde diethyl acetal (31.11 g, 0.20 mol) (Note 5) is added by syringe slowly, since the initial reaction is exothermic. The resulting solution is heated at 50°C with stirring for 2 hr. After the solution is cooled to room temperature, solvent is removed, first with a rotary evaporator and then with an oil pump overnight (Note 6). The residue is taken up in ethyl ether (Note 7), and filtered by suction through a 1 1/2-inch plug of Celite packed in a 250-mL coarse porosity, fritted Schlenk tube (Notes 8 and 9). The filtrate is collected in a 1-L, round-bottomed flask containing a magnetic stirring bar. The sodium chloride residue is washed several times with fresh ether until the washings are nearly colorless. The filter tube is removed and the flask is capped with a rubber septum and a nitrogen inlet; air is displaced by flushing the flask with nitrogen.

The solution is cooled to -78°C in a dry ice-acetone bath, and 38.19 g (0.23 mol) of tetrafluoroboric acid-diethyl ether complex (Note 10) is added dropwise by syringe over a 30-min period. The solution is allowed to warm to room temperature. The yellow precipitate is filtered off and collected in a 250-mL Schlenk tube, washed with ether, dried under a stream of nitrogen and finally under reduced pressure (oil pump). The bright yellow salt 1 weighs 43.2-57.6 g (60-80%) and may be used without further purification (Note 11). It may be stored indefinitely under nitrogen at 0°C without decomposition.

B. Dicarbonyl(cyclopentadienyl)(trans-3-methyl-2-vinylcyclohexanone)iron tetrafluoroborate **3**. Under a nitrogen atmosphere, cuprous iodide (Note 12) (24.76 g, 0.13 mol) and 150 mL of ether are placed in a 1-L, three-necked, round-bottomed flask containing a magnetic stirring bar, and cooled to 0°C in an ice-salt bath. First 172 mL (0.26 mol) of methyllithium in ether (Note 13) is added by syringe, then 12.60 g (0.13 mol) of 2-cyclohexen-1-one (Note 14) is added dropwise by syringe while the mixture is stirred at 0°C. A bright yellow precipitate forms immediately. After 15 min, 200 mL of tetrahydrofuran is added and the mixture is cooled to -78°C in a dry ice-acetone bath. While vigorous stirring and a strong flow of nitrogen are maintained, one septum is removed and 43.70 g (0.13 mol) of complex salt **1** is added at once. The septum is replaced and 50 mL of fresh tetrahydrofuran is used to wash solid **1** from the neck and sides of the reaction vessel. After 1 hr at -78°C, the mixture is allowed to warm to room temperature while stirring is continued. Stirring is halted to allow insoluble copper salts to settle leaving a red supernatant liquid. A 250-mL, coarse-frit, Schlenk filter is prepared with a Celite mat, topped with 2 inches of activity-IV neutral alumina (Note 15), which is further deactivated by washing in the Schlenk tube with 100 mL of diethyl ether. The supernatant solution is transferred to the Schlenk tube by cannula and filtered by suction into a 1-L, round-bottomed flask containing a magnetic stirring bar. The copper salts remaining in the reaction vessel are repeatedly washed with fresh ether until the filtered washings are nearly colorless. Removal of solvent from the filtrate leaves product **2** as a deep red oil (Note 16).

The oil is dissolved in 500 mL of diethyl ether under a nitrogen atmosphere, cooled to -78°C in a dry ice-acetone bath and 18 g (0.11 mol) of tetrafluoroboric acid-diethyl ether complex is added dropwise by syringe over a 30-min period, while the bath temperature is maintained at -78°C. The solution is allowed to warm to room temperature, and the powdery yellow solid is isolated by filtration through a 250-mL, coarse-frit, Schlenk filter tube. The product is washed several times with fresh ether and dried under a stream of nitrogen and then under reduced pressure (oil pump). The yield of salt **3** is 31-39.1 g (60-75%). The product may be used without further purification (Note 17) and may be stored under nitrogen at 0°C for several weeks with no observable decomposition (Note 18).

C. trans-3-Methyl-2-vinylcyclohexanone **4**. Compound **3** (31.5 g) and 25 mL of acetonitrile (0.47 mol, 6-fold excess) (Note 19) are placed in a 100-mL round-bottomed flask (Note 20) fitted with a magnetic stirring bar and reflux condenser. The mixture is heated to reflux for 2 hr under nitrogen, cooled to room temperature and slowly added to 300 mL of diethyl ether. The acetonitrile complex **5** precipitates as a bright yellow solid, and may be removed by suction filtration (Note 21).

The filtrate is washed three times with distilled water to remove excess acetonitrile and then dried over anhydrous magnesium sulfate. Filtration followed by removal of ether leaves the product **4** as a yellow oil (7.1-7.6 g, 64-70%), which may be further purified by short path or bulb-to-bulb distillation (bp 30°C at 0.1 mm) to a colorless liquid (Notes 22, 23).

2. Notes

1. All glassware, syringes and cannulae were routinely flame- or oven-dried and cooled under dry nitrogen.

2. In general, transfers of dry solvent or of solutions are made by 2-mm cannulae inserted through rubber septa capping delivery and receiver vessels. Transfer is made by positive nitrogen pressure applied through a hypodermic needle, while a second needle in the receiver vessel is employed as a vent. Cannulae are available from Hamilton Company, P. O. Box 10030, Reno, NV 89510.

3. Tetrahydrofuran is predried over potassium hydroxide pellets, degassed with nitrogen and freshly distilled under nitrogen from sodium benzophenone ketyl.[3]

4. Dicarbonyl(cyclopentadienyl)diiron can be readily made on a large scale from iron pentacarbonyl and dicyclopentadiene.[4] Alternatively it can be purchased from Alfa Products, Morton/Thiokol Inc. or from Strem Chemical Company.

5. Chloroacetaldehyde diethyl acetal was purchased from Aldrich Chemical Company, Inc. This substance is listed as an irritant. Proper care should be exercised when handling it.

6. It is important to stir the product continually to insure effective removal of tetrahydrofuran. If appreciable solvent remains, the vinyl ether complex 1 may not crystallize readily.

7. Ethyl ether was freshly distilled from sodium benzophenone ketyl.

8. Schlenk tubes were purchased from Ace Glass Company, catalogue #7761-36.

9. Filtration is smoothly accomplished by allowing the sodium chloride to settle and filtering the clear supernatant liquor in small portions, transferring the solution to the filtering tube by cannula. If the Celite should become clogged, the surface may be scraped clean, under a strong stream of nitrogen, using a long spatula, in order to increase the filtration rate.

10. Tetrafluoroboric acid - diethyl ether complex was purchased from Columbia Organics Chemical Company, Inc.

11. The salt may be reprecipitated by dissolution in methylene chloride containing a small amount of ethanol, followed by the addition of ether. The product shows the following spectra: IR (CH_2Cl_2) cm^{-1}: 2095, 2045, 1545; NMR (CD_3NO_2) δ: 1.40 (t, 3 H, Me), 2.73 (dd, 1 H, trans-CH=), 3.00 (dd, 1 H, cis-CH=), 4.37 (q, 2 H, OCH_2), 5.50 (s, 5 H, Cp), 7.92 (dd, 1 H, CHOEt). The checkers realized the higher yield only when the preparation was carried out on one-tenth the scale specified here.

12. Cuprous iodide was purchased from Fisher Scientific Company, and purified by recrystallization from saturated aqueous potassium iodide.[5,6]

13. Methyllithium was purchased from Aldrich Chemical Company, Inc., and standardized by double titration with benzoic acid in aqueous ethanol and with allyl chloride.[7]

14. 2-Cyclohexen-1-one was purchased from Aldrich Chemical Company, Inc., and purified by vacuum distillation.

15. Alumina was purchased from Woelm and brought to activity IV as directed.

16. As in the preparation of the vinyl ether complex **1**, it is very important to remove tetrahydrofuran effectively to promote facile crystallization of product. This is most easily done by removing most of the solvent with a rotary evaporator and then stirring the resulting oil under reduced pressure (oil pump) overnight. Mixing the oil with about 50 mL of ether followed by solvent removal under reduced pressure helped to facilitate removal of traces of tetrahydrofuran. Pure **2** may be obtained as a yellow crystalline solid, which decomposes on heating, by chromatographing the oil on basic alumina (activity IV), eluting with 5% ether in hexane. Compound **2** is characterized by the following spectra: IR(CH_2Cl_2) cm^{-1}: 2000, 1940, 1700; NMR ($CDCl_3$) δ: 1.17 (t, 6 H, CH_3), 1.43 (d, 2 H, $FeCH_2$), 1.65 (dd, 1 H, CHCO), 1.70-2.55 (m, 7 H, CH, CH_2), 3.41 (dq, 2 H, OCH_2), 3.76 (dt, 1 H, CHOEt), 4.87 (s, 5 H, Cp).

17. Compound **3** may be recrystallized by dissolution in a minimum volume of nitromethane at 0°C followed by the addition of diethyl ether. The crystalline material decomposes on heating. It is characterized by the following spectra: IR (CH_2CH_2) cm^{-1}: 2090, 2050, 1708; NMR (CD_3NO_2) δ: 1.1 (m, 3 H, CH_3), 1.4-2.6 (m, 8 H, CH, CH_2), 3.28 (d, 1 H, trans-CH_2=), 4.00 (d, 1 H, cis-CH_2=), 5.0 (m, 1 H, CH=), 5.65 (s, 5 H, Cp).

18. However, salt **3** is unstable at room temperature as a solid and in solution and rearranges to the isomeric complex in which the iron center is bound to the carbonyl group of the substituted cyclohexanone.

19. Acetonitrile is freshly distilled under nitrogen from calcium hydride.

20. Because of the ease with which product **4** isomerizes to the conjugated enone in the presence of base, it is imperative that the demetallation and subsequent purification steps be carried out in glassware that is free of basic residues.

21. An inert atmosphere is not necessary.

22. The considerable vapor pressure of **4** causes some loss during vacuum distillation. Although the color of the product is improved by distillation, both IR and NMR spectra of the product before and after distillation show little change.

23. Compound **4** is observed to darken on standing in air at room temperature for prolonged periods. Compound **4** is characterized by the following spectra: IR (neat) cm^{-1}: 1708; NMR ($CDCl_3$) δ: 0.99 (d,d 1 H, CH_3), 1.3-2.8 (m, 8 H, CH, CH_2), 4.98 (dd, 1 H, trans-CH_2=), 5.2 (dd, 1 H, cis-CH_2=), 5.76 (m, 1 H, CH=).

3. Discussion

Few reagents are available to the synthetic organic chemist which function as vinyl cation synthons. At present, these include α-phenylseleneoacetaldehyde,[8] α-silyl aldehydes and ketones,[9] α,β-epoxysilanes,[10] phenyl vinyl sulfoxide,[11] phenyl ethynyl sulfone,[12,13] phenyl 2-chlorovinyl sulfone,[13] and activated vinyl halides or ethynyl halides.[15] With the exception of the first two, the use of these reagents is confined to reactions with tertiary enolates or organocuprates.

The procedure given here illustrates the use of readily prepared organoiron complex 1 for the vinylation of a secondary enolate. This salt may be prepared on a large scale from readily available starting materials and can be stored at 0°C without decomposition. The closely related isopropenyl ethyl ether and cis-propenyl ethyl ether-iron complexes, 6 and 7, are similarly prepared from α-bromoacetone[16,17] and α-bromopropionaldehyde diethyl acetal[18] and have been used as isopropenylating[17] and trans-propenylating[19] reagents with enolates. Complex 8, derived from ethyl 3-bromopyruvate functions as an α-acrylic ester cation with enolates.[20,21]

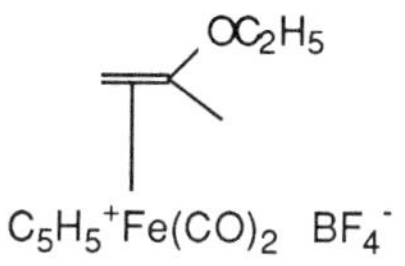

OC_2H_5 / $C_5H_5^+Fe(CO)_2$ BF_4^-

6

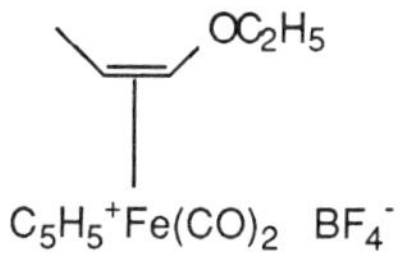

OC_2H_5 / $C_5H_5^+Fe(CO)_2$ BF_4^-

7

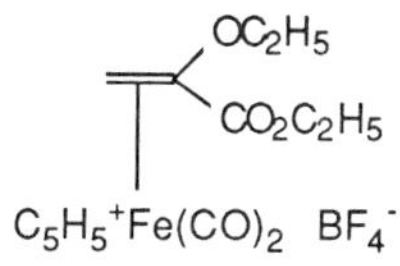

OC_2H_5 / $CO_2C_2H_5$ / $C_5H_5^+Fe(CO)_2$ BF_4^-

8

Because of their high reactivity, these complex salts react rapidly and regiospecifically, at low temperature, with a number of carbon and heteroatomic nucleophiles, including thiols, amines, and alcohols.[22] Finally, exposure of the double bond takes place under particularly mild conditions so that isomerization of the β,γ-unsaturated carbonyl system may be avoided. The present scope of reactions with these vinyl cation synthons is summarized in Table I.

TABLE

VINYLATION OF ENOLATES WITH VINYL CATION EQUIVALENTS

Enolate	Enol Ether Complex	Vinylated Product	Yield %
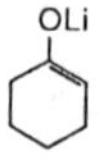 OLi	1	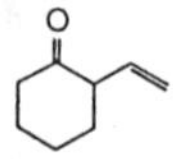 O	80
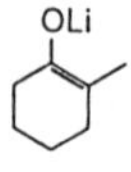 OLi	1	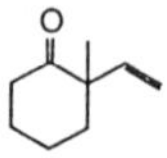 O	80
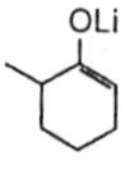 OLi	1	 O + O 1 : 1	80
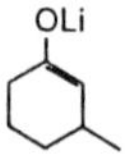 OLi	1	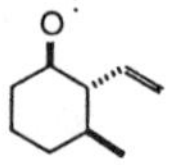 O	47[a]
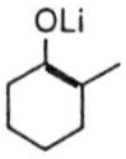 OLi	7	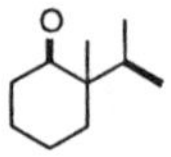 O	72
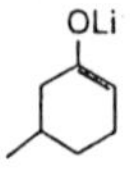 OLi	7	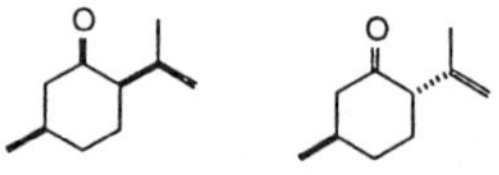 O O 2 : 5	53
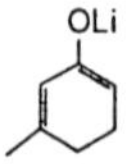 OLi	7	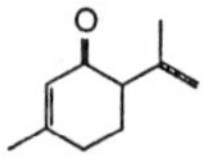 O	91
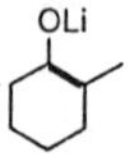 OLi	6	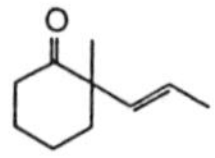 O	72
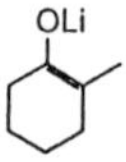 OLi	8	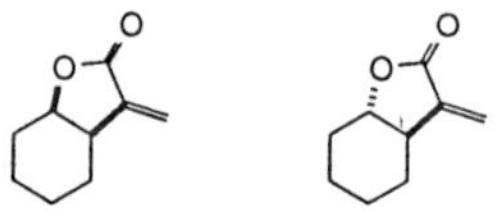 O O O O	44[b] 37[c]
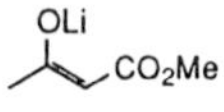 OLi CO_2Me	8	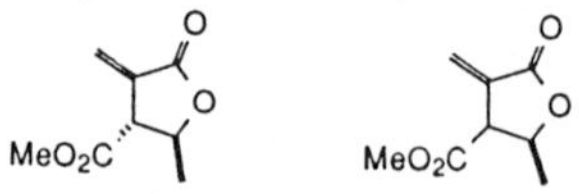 O O MeO_2C 9 O O MeO_2C 1	65[b]
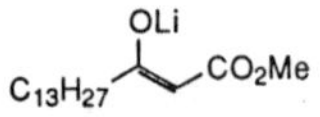 OLi $C_{13}H_{27}$ CO_2Me	8	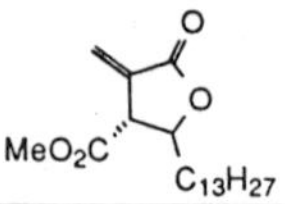 O O MeO_2C $C_{13}H_{27}$	(15)[d]

[a]From cyclohexenone.

[b]By reduction of the initial condensation product with L-selectride.

[c]By reduction of the initial condensation product with sodium borohydride.

[d]By reduction of the initial condensation product with $LiAlH_4$ at -78^{0}C.

1. Present address, General Electric Co., Corporate Research and Development, P.O. Box 8, Schenectady, NY 12301.

2. Department of Chemistry, Brandeis University, Waltham, MA 02254.

3. Gordon, A. J.; Ford, R. A. "The Chemist's Companion"; Wiley: New York, 1972; p. 439.

4. Eisch, J. J.; King, R. B., Eds. In "Organometallic Syntheses"; Academic Press: New York, 1965; Vol. 1, p. 114.

5. House, H. O.; Respess, W. L.; Whitesides, G. M. *J. Org. Chem.* **1966**, *31*, 3128.

6. To dry cuprous iodide, see Gmelins Handbuch der Anorganischen Chemie. Kupfer, Teil B, Lieferung 1, 390.

7. Gilman, H.; Cartledge, F. K. *J. Organomet. Chem.* **1964**, *2*, 447.

8. Clive, D. L. J.; Russel, C. G.; Suri, S. C. *J. Org. Chem.* **1982**, *47*, 1632; Kowalski, C, J.; Dung, J.-S. *J. Am. Chem. Soc.* **1980**, *102*, 7950.

9. Hudrlik, P. F.; Kulkarni, A. K. *J. Am. Chem. Soc.* **1981**, *103*, 6251; Hudrlik, P. F.; Peterson, D. *Tetrahedron Lett.* **1972**, 1785; Hudrlik, P. F.; Peterson, D. *Tetrahedron Lett.* **1974**, 1133; Hudrlik, P. F.; Peterson, D. *J. Am. Chem. Soc.* **1975**, *97*, 1464; Ruden, R. A.; Gaffney, B. L. *Synth. Commun.* **1975**, *5*, 15; Utimoto, K.; Obayashi, M.; Nozaki, H. *J. Org. Chem.* **1976**, *41*, 2940; Obayaski, M.; Utimoto, K.; Nozaki, H. *Tetrahedron Lett.* **1977**, 1807.

10. Hudrlik, P. F.; Peterson, D.; Rona, R. J. *J. Org. Chem.* **1975**, *40*, 2263; Eisch, J. J.; Galle, J. E. *J. Org. Chem.* **1976**, *41*, 2615.

11. Koppel, G. A.; Kinnick, M. D. *J. Chem. Soc., Chem. Commun.* **1975**, 473; Bruhn, J.; Heimgartner, H.; Schmid. H. *Helv. Chim. Acta* **1979**, *62*, 2630.

12. Steglich, W.; Wegmann, H. *Synthesis* **1980**, 481.

13. Metcalf, B. W.; Bonilavri, E. *J. Chem. Soc., Chem. Commun.* **1978**, 914.

14. Bey, P.; Vevert, J. P. *J. Org. Chem.* **1980**, *45*, 3249; Millard, A. A.; Rathke, M. W. *J. Am. Chem. Soc.* **1977**, *99*, 4833.

15. Kende, A. S.; Benechie, M.; Curran, D. P.; Fludzinski, P.; Swenson, W.; Clardy, J. *Tetrahedron Lett.* **1979**, 4513; Kende, A. S.; Fludzinski, P. *Tetrahedron Lett.* **1982**, *23*, 2373.

16. Abram, T. S.; Baker, R. *Synth. React. Inorg. Met.-Org. Chem.* **1979**, 471; *Chem. Abstr.* **1979**, *91*, 193392q.

17. Chang, T. C. T.; Rosenblum, M. *J. Org. Chem.* **1981**, *47*, 4103.

18. Cutler, A.; Raghu, S.; Rosenblum, M. *J. Organomet. Chem.* **1974**, *77*, 381.

19. Chang, T. C. T.; Rosenblum, M.; Samuels, S. B. *J. Am. Chem. Soc.* **1980**, *102*, 5930.

20. Chang, T. C. T.; Rosenblum, M. *J. Org. Chem.* **1981**, *46*, 4626.

21. Chang, T. C. T.; Rosenblum, M. *Tetrahedron Lett.* **1983**, *24*, 695.

22. Chang, T. C. T.; Foxman, B. M.; Rosenblum, M.; Stockman, C. *J. Am. Chem. Soc.* **1981**, *103*, 7361.

Appendix

Chemical Abstracts Nomenclature (Collective Index Number); (Registry Number)

Dicarbonyl(cyclopentadienyl) (ethyl vinyl ether)iron tetrafluoroborate: Iron(1+), dicarbonyl(η^5-2,4-cyclopentadien-1-yl)[(1,2-η)-ethoxyethene]-, tetrafluoroborate(1-) (10); (75182-42-2)

Dicarbonyl(cyclopentadienyl) iron dimer: Iron, di-μ-carbonyldicarbonyldi-π-cyclopentadienyldi-, (Fe-Fe) (8); Iron, di-μ-carbonyldicarbonylbis(η^5-2,4-cyclopentadien-1-yl)di-, (Fe-Fe) (9); (12154-95-9)

Sodium dicarbonyl(cyclopentadienyl)ferrate: Ferrate (1-), dicarbonyl-π-cyclopentadienyl-, sodium (8); Ferrate (1-), dicarbonyl(η^5-2,4-cyclopentadien-1-yl)-, sodium (9); (12152-20-4)

Chloroacetaldehyde diethyl acetal: Acetaldehyde, chloro-, diethyl acetal (8); Ethane, 2-chloro-1,1-diethoxy- (9); (621-62-5)

Tetrafluoroboric acid - diethyl ether complex: Borate (1-), tetrafluoro-, hydrogen, compd. with 1,1'-oxybis[ethane] (1:1) (10); (67969-82-8)

2-Cyclohexen-1-one (8,9); (930-68-7)

PREPARATION OF tert-BUTYL ACETOTHIOACETATE AND USE IN THE SYNTHESIS OF 3-ACETYL-4-HYDROXY-5,5-DIMETHYLFURAN-2(5H)-ONE

(Acetoacetic acid, 1-thio-, S-tert-butyl ester)

A. $(CH_3)_3CSH$ + NaH ⟶ $(CH_3)_3CS^- Na^+$

$(CH_3)_3CS^- Na^+$ + [diketene] ⟶ $(CH_3)_3CSC(O)CH_2C(O)CH_3$

B. $(CH_3)_3CSC(O)CH_2C(O)CH_3$ + $(CH_3)_2C(OH)CO_2Me$ —$AgOCOCF_3$⟶ $MeO_2CC(CH_3)_2OC(O)CH_2C(O)CH_3$

C. $MeO_2CC(CH_3)_2OC(O)CH_2C(O)CH_3$ —$Bu_4N^+ F^-$⟶ 3-acetyl-4-hydroxy-5,5-dimethylfuran-2(5H)-one

Submitted by Christina M. J. Fox and Steven V. Ley.[1]

Checked by Anura P. Dantanarayana and James D. White.

1. Procedure

Caution! 2-Methylpropane-2-thiol should be handled in an efficient fume hood because of its odor.

A. S-tert-Butyl 3-oxobutanthioate. A dry, 2-L, three-necked, round-bottomed flask (Note 1) fitted with a 100-mL pressure-equalizing dropping funnel, thermometer, magnetic stirrer bar and an argon inlet, is charged with

9.8 g (0.24 mol) of sodium hydride as a 60% dispersion in oil (Note 2). The system is flushed with, and kept under, dry argon. The sodium hydride is washed with two 40-mL portions of sodium-dried pentane and the system is purged with dry argon to remove traces of pentane. To the flask is then added 900 mL of dry tetrahydrofuran (Note 3).

The flask is cooled in an ice-salt bath to -5°C and a solution of 20 g (25 mL, 0.22 mol) of 2-methylpropane-2-thiol (Note 2) in 20 mL of dry tetrahydrofuran is added at such a rate as to maintain a steady evolution of hydrogen. The slightly exothermic reaction causes the temperature to rise to 0°C and the colorless solution is stirred at this temperature for 15 min to ensure complete formation of the thiolate. The reaction mixture is then recooled to -5°C and 20.3 g (18.8 mL, 0.24 mol) of diketene (Note 2) is added over 15 min to give a yellow-green solution. The cooling bath is removed and the solution allowed to warm to room temperature.

The reaction is quenched and excess sodium hydride is destroyed by careful addition of 300 mL of saturated ammonium chloride solution. The two-phase mixture is transferred to a 2-L separatory funnel charged with 400 mL of ether. The layers are separated and the organic phase is washed with 300-mL portions of water, saturated sodium bicarbonate solution and brine. The aqueous washes are re-extracted with a 400-mL portion of ether and the combined organic layers are dried over anhydrous sodium sulfate. The solvent is removed with a rotary evaporator to give the crude product as a deep red oil. Bulb-to-bulb distillation (Note 4) at 95-100°C (0.9 mm) gives 22 g (57%) of S-tert-butyl 3-oxobutanthioate as a colorless oil (Note 5).

B. 1-Carbomethoxy-1-methylethyl 3-oxobutanoate. A 500-mL, round-bottomed flask equipped with a magnetic stirrer bar is charged with 10 g (0.085 mol) of methyl 2-hydroxyisobutyrate (Note 2), 17.7 g (0.102 mol) of S-tert-butyl 3-oxobutanthioate and 250 mL of dry tetrahydrofuran. The flask is placed in the dark and 22.5 g (0.102 mol) of freshly prepared silver(I) trifluoroacetate (Note 6) is added in two portions. The resulting dark brown suspension is stirred for 15 min (Note 7) and then concentrated to approximately 50 mL with a rotary evaporator. The concentrated mixture is diluted with 200 mL of hexane and the resulting orange-brown precipitate is removed by filtration. The filtered solid is washed with two 50-mL portions of hexane and the combined filtrate and washings concentrated with a rotary evaporator to give an orange-brown oil.

The crude product is chromatographed on 350 g of silica (Note 8) using 1:1 ether-petroleum ether (40-60) as eluant. The chromatography is monitored by TLC (Note 9) and the appropriate fractions are combined. Removal of the solvent with a rotary evaporator gives a pale orange oil (Note 10) which was further purified by distillation to give 11.7 g (68%) of the O-ester, bp 69-72°C (0.2 mm) (Note 11).

C. 3-Acetyl-4-hydroxy-5,5-dimethylfuran-2(5H)-one. A 100-mL, round-bottomed flask equipped with a 50-mL pressure-equalizing dropping funnel and a magnetic stirrer bar is charged with 5 g (0.025 mol) of the acetoacetate and 37 mL (0.037 mol) of tetrabutylammonium fluoride (1 M solution in THF) (Note 2) is added over 5 min. The resulting solution is stirred vigorously for 3 hr (Note 12) and then transferred to a 250-mL separatory funnel containing 50 mL of 6 M hydrochloric acid. The acidified mixture is extracted with three 30-mL portions of ether, each extract being washed with 10 mL of brine. The combined organic extracts are dried over anhydrous sodium sulfate and concentrated with a rotary evaporator to give 4.3 g of the crude tetronic acid as a yellow solid. Recrystallization from 25 mL of hot 5% ether - petroleum ether gives 1.9 g of the tetronic acid as pale yellow plates, mp 66-67°C (lit[2]: mp 64-65°C) (Note 13). Concentration of the mother liquor affords a second crop of 0.4 g, mp 63-65°C, giving a combined yield of 54%.

2. Notes

1. The apparatus was oven dried, assembled while hot and cooled under a stream of dry argon.

2. Sodium hydride, 2-methylpropane-2-thiol, diketene, methyl 2-hydroxyisobutyrate and tetrabutylammonium fluoride were purchased from Aldrich Chemical Company, Inc. Diketene was distilled prior to use to remove polymeric species.

3. Tetrahydrofuran was refluxed over and distilled from sodium/benzophenone immediately prior to use.

4. A Kugelrohr apparatus was used for the distillation. The reported temperature is the oven temperature.

5. The checkers found that the yield of this material was substantially higher (83%) when the reaction was conducted at 1/4 scale. Spectral properties of the product[3] are as follows: IR (neat) cm^{-1}: 1712, 1676, 1621; ^{1}H NMR (60 MHz, $CDCl_3$) δ: 1.5 (s, 9 H, $(CH_3)_3C$), 2.3 (s, 3 H, $COCH_3$), 3.6 (s, 2 H, $COCH_2CO$), 5.3 (s, COCH=C(OH)).

6. Silver(I) trifluoroacetate may be obtained commercially but it is recommended that it be freshly prepared.[4] Trifluoroacetic acid (18 mL, 0.24 mol) is added to silver(I) oxide (0.12 mol), freshly precipitated from silver nitrate (20 g, 0.12 mol) and sodium hydroxide (4.7 g, 0.12 mol) in water (30

mL). The solution is filtered and evaporated to dryness under reduced pressure. The crude product is purified by dissolving it in ether (150 mL), filtering through decolorizing charcoal and evaporation to give the product as a white crystalline solid (19.3 g, 74%).

7. The time reported represents the average reaction time. The reaction can be followed by TLC, visualizing with iodine and 10% phosphomolybdic acid in ethanol followed by heating on a hot plate.

8. Merck Kieselgel 60 silica gel (230-400 mesh) was used.

9. Merck precoated silica gel 60 F-254 plates were used, visualizing with iodine.

10. In some cases, the product is contaminated with a yellow solid even after chromatography. This is removed prior to distillation by filtering through a short pad of Celite.

11. The spectral properties of the product[2] are as follows: IR (neat) cm^{-1}: 1745, 1720; ^{1}H NMR δ (90 MHz, $CDCl_3$): 1.57 (s, 6 H, $(CH_3)_2C$), 2.28 (s, 3 H, $COCH_3$), 3.44 (s, 2 H, $COCH_2CO$), 3.72 (s, 3 H, CO_2CH_3).

12. The reaction is monitored by TLC and quenched when starting material has been consumed.

13. Spectral properties of the product[2] are as follows: IR (KBr) cm^{-1}: 1758, 1685, 1610; ^{1}H NMR (60 MHz, $CDCl_3$) δ: 1.50 and 1.51 (2 s, 6 H, $C(CH_3)_2$), 2.5 (s, 3 H, $COCH_3$), 9.25 (br s, 1 H, OH).

3. Discussion

Selective alkylation of β-keto esters via either anions or dianions is an important synthetic transformation.[5] Equally, thioesters may be trans-esterified in the presence of thiophilic metal cations.[6] These two features can be usefully combined in one substrate, tert-butyl acetothioacetate, the subject of this *Organic Syntheses* procedure.

Alkylation at the 2-position can be achieved by formation of the anion with sodium hydride in 1,2-dimethoxyethane (DME) at 0°C followed by reaction with an alkyl halide at room temperature. Alternatively, selective alkylation at C-4 involves sequential treatment with sodium hydride (at -10°C) and butyllithium in DME (at -40°C) to form the dianion, followed by kinetic alkylation with an alkyl halide (or carbonyl compound).[7]

$(CH_3)_3CS$ O O

$(CH_3)_3CS$ O O R

$(CH_3)_3CS$ O O R

The choice of DME as solvent in these reactions is important as other ether solvents are much less successful and lead to unwanted side products.

Transesterification of the resulting alkylated β-keto thioesters to the corresponding oxo esters is readily achieved using alcohols under various metal catalysis.[6]

The alcohols used may also contain fairly sensitive functional groups, e.g., esters, halides, silyl ethers, etc. In this work, therefore, tert-butyl acetothioacetate is behaving as a synthetic equivalent to diketene. When this methodology is used, it is possible to devise very short syntheses of acyl tetronic acids[7] and novel macrocyclic structures.[8]

1. Department of Chemistry, Imperial College, London SW7 2AY.

2. Lacey, R. N. *J. Chem. Soc.* **1954**, 832.

3. (a) Wilson, G. E., Jr.; Hess, A. *J. Org. Chem.* **1980**, *45*, 2766 and references therein; (b) Motoki, S.; Sato, T. *Bull. Chem. Soc. Jpn.* **1969**, *42*, 1322.

4. Janssen, D. E.; Wilson, C. V. *Org. Synth., Collect. Vol. 4* **1963**, 547.

5. Weiler, L. *J. Am. Chem. Soc.* **1970**, *92*, 6702; Huckin, S. N.; Weiler, L. *J. Am. Chem. Soc.* **1974**, *96*, 1082; Sum, F. W.; Weiler, L. *Tetrahedron, Suppl.* **1981**, *(9)*, 303-317; *Chem. Abstr.* **1981**, *95*, 150929k.

6. Masamune, S.; Hayase, Y.; Schilling, W.; Chan, W. K.; Bates, G. S. *J. Am. Chem. Soc.* **1977**, *99*, 6756.

7. Booth, P. M.; Fox, C. M. J.; Ley, S. V. *J. Chem. Soc., Perkin Trans. 1* **1987**, 121.

8. Fox, C. M. J.; Ley, S. V.; Slawin, A. M. Z.; Williams, D. J. *J. Chem. Soc., Chem. Commun.* **1985**, 1805.

Appendix

Chemical Abstracts Nomenclature (Collective Index Number); (Registry Number)

tert-Butyl acetothioacetate: Acetoacetic acid, 1-thio-, S-tert-butyl ester (8,9); (15925-47-0)

2-Methylpropane-2-thiol: 2-Propanethiol, 2-methyl- (8,9); (75-66-1)

Sodium hydride (8,9); (7646-69-7)

Diketene: 2-Oxetanone, 4-methylene- (8,9); (674-82-8)

Methyl 2-hydroxyisobutyrate: Lactic acid, 2-methyl-, methyl ester (8); Propanoic acid, 2-hydroxy-2-methyl-, methyl ester (9); (2110-78-3)

Silver(I) trifluoroacetate: Acetic acid, trifluoro-, silver (1 + salt) (8,9); (2966-50-9)

Tetrabutylammonium fluoride: Ammonium, tetrabutyl-, fluoride (8); 1-Butanaminium, N,N,N-tributyl-, fluoride (9); (429-41-4)

Trifluoroacetic acid: Acetic acid, trifluoro- (8,9); (76-05-1)

Silver(I) oxide: Silver oxide (8); Silver oxide (Ag_2O) (9); (20667-12-3)

KETONES FROM CARBOXYLIC ACIDS AND GRIGNARD REAGENTS:

METHYL 6-OXODECANOATE

(Decanoic acid, 6-oxo-, methyl ester)

$$HO_2C(CH_2)_4CO_2Me + Me_2C{=}C(Cl)NMe_2 \longrightarrow ClCO(CH_2)_4CO_2Me$$

$$ClCO(CH_2)_4CO_2Me + CH_3(CH_2)_3MgBr \xrightarrow{\text{cat. CuI}} CH_3(CH_2)_3CO(CH_2)_4CO_2Me$$

Submitted by Tamotsu Fujisawa and Toshio Sato.[1]

Checked by Cynthia Smith and Andrew S. Kende.

1. Procedure

A 500-mL, three-necked, round-bottomed flask equipped with a magnetic stirring bar, thermometer, an addition funnel for solids (Note 1), and a rubber septum is flushed with nitrogen. The flask is charged with 50 mL of dichloromethane (Note 2) and 6.92 g (0.052 mol) of 1-chloro-N,N,2-trimethylpropenylamine (Note 3). The solution is stirred and cooled in an ice bath and 8.01 g (0.050 mol) of adipic acid monomethyl ester (Note 4) is added slowly by means of a syringe over 10 min. After the addition is complete, the cooling bath is removed and the contents of the flask are stirred for 30 min at room temperature (Note 5). The flask is cooled in an ice-salt bath to -15°C. Then 100 mL of tetrahydrofuran (Note 6) and 0.48 g (0.0025 mol) of copper(I) iodide (Note 7) are added to the flask through the septum and the funnel, respectively. To this stirred mixture is added 50.5 mL (0.052 mol) of a 1.03 M solution of butylmagnesium bromide (Note 8) in tetrahydrofuran over 1 hr using a syringe pump, while the internal temperature is maintained below -10°C. The reaction mixture is stirred for an additional hour at -15°C. After 100 mL of 2 M hydrochloric acid solution has been poured into the flask in one portion, the mixture is transferred to a separatory funnel and the organic layer is separated. The aqueous layer is extracted with two 100-mL portions of hexane. The combined organic extracts are washed with five 100-mL portions of 2 M hydrochloric acid solution (Note 9), 100 mL of 5% sodium thiosulfate solution, two 100-mL portions of saturated sodium bicarbonate solution, and 100 mL of brine, dried over anhydrous sodium sulfate, and filtered. The solvent is evaporated under reduced pressure and the residual liquid is distilled with a short-necked Claisen distillation flask. After separation of a small forerun (<0.3 g) (Note 10), 8.53-8.67 g (85-86%) of methyl 6-oxodecanoate is collected, bp 106-110°C (2.8 mm) (Note 11).

2. Notes

1. A simple bent glass tube is useful as an addition funnel for copper(I) iodide.

2. Dichloromethane was distilled over calcium hydride, and stored over molecular sieves 4 Å.

3. N,N-Dimethylisobutyramide (Gavrilov, N.; Koperina, A.; Klutcharova, M. *Bull. Soc. Chim. France* **1945**, *12*, 773) was converted to 1-chloro-N,N,2-trimethylpropenylamine according to the procedure of *Org. Synth.* **1979**, *59*, 26, in 61% yield, bp 118-121°C. Freshly-distilled oxalyl chloride was used instead of phosgene. The propenylamine should be handled carefully in a syringe to avoid its rapid hydrolysis by moisture.

4. Adipic acid monomethyl ester was purchased from Nakarai Chemicals, or Aldrich Chemical Company, Inc. and distilled before use, bp 155-158°C (7 mm).

5. In a separate experiment, formation of adipic acid monomethyl ester monochloride was observed.[2]

6. Tetrahydrofuran was freshly distilled from the sodium ketyl of benzophenone.

7. Copper(I) iodide purchased from Wako Chemicals was used without purification.

8. Butylmagnesium bromide was prepared from magnesium and butyl bromide in tetrahydrofuran at room temperature by a standard procedure (*Org. Synth.* **1978**, *58*, 127), and titrated by the procedure of Eastham, et al.[3]

9. The organic extracts must be washed four or five times to remove N,N-dimethylisobutyramide.

10. The forerun consisted of N,N-dimethylisobutyramide, other by-products, and methyl 6-oxodecanoate.

11. The reported physical constants are bp 149°C (13.5 mm),[4] 97-103°C (3.5 mm),[5] n_D^{20} 1.4377,[4] n_D^{25} 1.4376.[5] Gas chromatographic analysis of the product using a 3 mm x 1-m stainless steel column, 15% SE-30 on 60-80 mesh chromosorb W (AW), 150°C, 50 mL of nitrogen per min indicated a purity of 99.6% (the retention time is 6.9 min). The spectral properties of the product are as follows: IR (liquid film) cm^{-1}: 2960, 2870, 1740, 1714, 1454, 1435, 1415, 1370, 1200, and 1175; ^1H NMR (60 MHz, CCl_4) δ: 0.9 (t, 3 H, J = 7, CH_3), 1.06-1.86 (m, 8 H, CH_2), 2.06-2.56 (m, 6 H, $CH_2C=O$), 3.60 (s, 3 H, OCH_3).

3. Discussion

The direct coupling of Grignard reagents with carboxylic acids is not generally useful for ketone synthesis because of the accompanying formation of tertiary alcohols. An exception is the recently-published method using a nickel catalyst.[6] In order to accomplish such a chemoselective ketone synthesis, the method of activation of carboxylic acid in situ is important, and several activating reagents have been proposed for the purpose, such as a bulky acyl chloride,[7] dichlorotriphenylphosphorane,[8] or N,N-diphenyl-p-methoxyphenylchloromethylenammonium chloride,[9] which react with carboxylic acids to produce mixed anhydrides, carboxyphosphonium salts, or carboxymethylenammonium salts, respectively.

The present procedure, reported earlier by the submitters,[10] illustrates a general method for ketone synthesis in a one-pot operation using an α-chloroenamine as a condensation reagent. 1-Chloro-N,N,2-trimethylpropenylamine reacts with carboxylic acids to produce the corresponding acyl chlorides[2] which instantaneously couple with Grignard reagents in the presence of a copper catalyst to give ketones. The utility of the procedure is as follows: (1) an equimolecular amount of Grignard reagent is sufficient to complete the reaction of carboxylic acid; and (2) the exceptionally high chemoselectivity of the reaction tolerates various kinds of functional groups such as nitrile, halide, ester and even ketone.[10]

1. Chemistry Department of Resources, Mie University, Tsu Mie 514, Japan.
2. Devos, A.; Remion, J.; Frisque-Hesbain, A.-M.; Colens, A.; Ghosez, L. *J. Chem. Soc., Chem. Commun.* **1979**, 1180.
3. Watson, S. C.; Eastham, J. F. *J. Organomet. Chem.* **1967**, *9*, 165.

4. Brockman, J. A., Jr.; Fabio, P. F. *J. Am. Chem. Soc.* **1957**, *79*, 5027.

5. Holmquist, H. E.; Rothrock, H. S.; Theobald, C. W.; Englund, B. E. *J. Am. Chem. Soc.* **1956**, *78*, 5339.

6. Fiandanese, V.; Marchese, G.; Ronzini, L. *Tetrahedron Lett.* **1983**, *24*, 3677.

7. Araki, M.; Mukaiyama, T. *Chem. Lett.* **1974**, 663.

8. Fujisawa, T.; Iida, S.; Uehara, H.; Sato, T. *Chem. Lett.* **1983**, 1267.

9. Fujisawa, T.; Mori, T.; Sato, T. *Tetrahedron Lett.* **1982**, *23*, 5059.

10. Fujisawa, T.; Mori, T.; Higuchi, K.; Sato, T. *Chem. Lett.* **1983**, 1791.

Appendix

Chemical Abstracts Nomenclature (Collective Index Number); (Registry Number)

Methyl 6-oxodecanoate: Decanoic acid, 6-oxo-, methyl ester (10); (61820-00-6)

1-Chloro-N,N,2-trimethylpropenylamine: Propenylamine, 1-chloro-N,N,2-trimethyl- (8); 1-Propen-1-amine, 1-chloro-N,N,2-trimethyl- (9); (26189-59-3)

Adipic acid monomethyl ester (8); Hexanedioic acid, monomethyl ester (9); (627-91-8)

REDUCTION OF CARBOXYLIC ACIDS TO ALDEHYDES: 6-OXODECANAL

(Decanal, 6-oxo-)

A. $HC(O)NMe_2$ + $(COCl)_2$ ⟶ $ClCH{=}N^+Me_2$ Cl^- + CO + CO_2

B. 6-oxodecanoic acid + $ClCH{=}N^+Me_2$ Cl^- ⟶ 6-oxodecanoyl chloride

6-oxodecanoyl chloride + $LiAlH(OCMe_3)_3$ ⟶ 6-oxodecanal (CHO)

Submitted by Tamotsu Fujisawa and Toshio Sato.[1]

Checked by Cynthia Smith and Andrew S. Kende.

1. Procedure

Caution! Oxalyl chloride is toxic. This preparation should be carried out in a well-ventilated hood.

A. *N,N-Dimethylchloromethylenammonium chloride.* A 500-mL, three-necked, round-bottomed flask is equipped with a magnetic stirring bar, thermometer (Note 1), and a three way stopcock fitted with a drying tube containing anhydrous calcium chloride and a rubber septum. The flask is charged with 50 mL of dichloromethane (Note 2) and 3.07 g (0.042 mol) of N,N-dimethylformamide (Note 3) added through the septum from a syringe, and cooled in an ice bath. To the cooled mixture is slowly added 5.23 mL (0.06 mol) of oxalyl chloride (Note 4) by means of a syringe. The addition is accompanied by gas evolution and formation of a white precipitate. The reaction mixture is stirred for an additional hour at 0°C. Excess oxalyl chloride and solvent are removed under reduced pressure by first using a water aspirator and then a rotary pump at room temperature through the drying tube. The white solid remaining in the flask is N,N-dimethylchloromethylenammonium chloride, which is used directly in Part B.

B. 6-Oxodecanal. The drying tube is removed and the flask is flushed with nitrogen. A nitrogen atmosphere is maintained throughout the subsequent reaction. A dropping funnel is attached and charged with 7.45 g (0.04 mol) of 6-oxodecanoic acid (Note 5), 3.32 g of pyridine (Note 6) and 80 mL of tetrahydrofuran (Note 7), which are mixed well by shaking. The flask is charged with 45 mL of acetonitrile (Note 8) and 80 mL of tetrahydrofuran, and cooled (methanol-liquid nitrogen) to -30°C. The contents of the funnel are added to the flask at -30°C over 30 min. The reaction mixture is stirred at -30°C for an additional hour and at -20°C for 30 min. After the mixture is cooled to -90°C (Note 9), 34 mL (0.046 mol) of a 1.35 M solution of lithium tri(tert-butoxy)aluminum hydride in tetrahydrofuran (Note 10) is injected through the septum by means of a syringe over 30 min, while the internal temperature is kept below -85°C. Stirring is continued for an additional 30 min at -90°C. To the flask is added 50 mL of 2 M hydrochloric acid solution, and the cooling bath is immediately removed. The organic layer is separated and the aqueous layer is extracted with three 50-mL portions of ether. The combined organic extracts are washed with two 50-mL portions of saturated sodium hydrogen carbonate solution and 50 mL of brine, dried over anhydrous sodium sulfate, and filtered. The solvent is removed with a rotary evaporator and the residual liquid is distilled under reduced pressure to yield 5.78-6.35 g (85-93%) of 6-oxodecanal as a fragrant liquid, bp 85-90°C (1.4 mm) (Note 11).

2. Notes

1. The thermometer must be able to measure temperatures as low as -90°C.

2. Dichloromethane was distilled from calcium hydride, and stored over molecular sieves 4 Å.

3. N,N-Dimethylformamide was distilled under reduced pressure, bp 45-47°C (20 mm), and stored over molecular sieves 4 Å.

4. Oxalyl chloride purchased from Wako Chemicals was used without purification. The checkers found that oxalyl chloride, purchased from Aldrich Chemical Company, Inc., gives better yields if freshly distilled.

5. 6-Oxodecanoic acid was obtained by hydrolysis of methyl 6-oxodecanoate prepared by the method of *Org. Synth.*[2] as follows: Twenty grams (0.100 mol) of methyl 6-oxodecanoate was treated with 200 mL of 1 M potassium hydroxide solution at room temperature overnight. The alkaline solution was washed with two 50-mL portions of ether, and acidified with 50 mL of 6 M hydrochloric acid solution at 0°C. The acidic layer was extracted with three 100-mL portions of ether. The ethereal extracts were dried over sodium sulfate, and filtered. Removal of the solvent under reduced pressure and recrystallization of the residual white solid from hexane gave 17.85 g (96%) of 6-oxodecanoic acid, mp 45.0-45.5°C (lit.[3] mp 45-46°C).

6. Pyridine was distilled from calcium hydride, and stored over molecular sieves 4 Å.

7. Tetrahydrofuran was freshly distilled from the sodium ketyl of benzophenone.

8. Acetonitrile was distilled from calcium hydride, and stored over molecular sieves 4 Å.

9. The checkers used a 1:1 methanol:ethanol/liquid nitrogen bath.

10. Lithium tri(tert-butoxy)aluminum hydride was purchased from Kanto Chemicals or Aldrich Chemical Company, Inc.

11. GLC analysis of the product using a 3 mm x 1 m stainless steel column, 15% SE-30 on chromosorb W (AW), 60-80 mesh, 150°C, 50 mL of nitrogen per min showed a purity of 99.2% (the retention time is 4.0 min). The spectral properties of the product are as follows: IR (liquid film) cm^{-1}: 2950, 2930, 2870, 2720, 1710, 1455, 1410, and 1370; ^{1}H NMR (60 MHz, CCl_4) δ: 0.87 (t, 3 H, J = 7, CH_3), 1.1-1.9 (m, 8 H, CH_2), 2.1-2.6 (m, 6 H, $CH_2C=O$), 9.73 (t, 1 H, J = 1.4, H-C=O).

3. Discussion

Various reagents have been suggested for the conversion of carboxylic acids into aldehydes, such as modified aluminum hydride reagents,[4] Grignard reagents catalyzed by dichlorobis(π-cyclopentadienyl)titanium,[5] lithium in methylamine,[6] and boron hydride reagents.[7] However, these reagents have some drawbacks in availability, lack of chemoselectivity due to the high reactivity of the reagents, and isolation of products.

The present procedure, a modified one reported earlier by the submitters, illustrates a general method of aldehyde synthesis from carboxylic acids in a one-pot operation using the readily available N,N-dimethylchloromethylenammonium chloride.[8] Strong activation of carboxylic acids by the iminium salt via the carboxymethylenammonium salt[9] and a weak reducing reagent, lithium tri(tert-butoxy)aluminum hydride, achieve the chemoselective reduction of carboxylic acids to aldehydes. The present procedure has several advantages: (1) easy availability of the reagents, (2) use of a slight excess of the hydride reagent, (3) high yields of both aliphatic and aromatic aldehydes, (4) high chemoselectivity, which tolerates nitrile, ester, halide, olefin, and even ketone, and (5) easy isolation of the product.

1. Chemistry Department of Resources, Mie University, Tsu, Mie 514, Japan.
2. Fujisawa, T.; Sato, T. *Org. Synth.* **1988**, *66*, 116.
3. Holmquist, H. E.; Rothrock, H. S.; Theobald, C. W.; Englund, B. E. *J. Am. Chem. Soc.* **1956**, *78*, 5339.
4. Muraki, M.; Mukaiyama, T. *Chem. Lett.* **1974**, 1447; Zakharkin, L. I.; Khorlina, I. M. *Zh. Obshch. Khim.* **1964**, *34*, 1029; *Chem. Abstr.* **1964**, *60*, 15724f; Hubert, T. D.; Eyman, D. P.; Wiemer, D. F. *J. Org. Chem.* **1984**, *49*, 2279.
5. Sato, F.; Jinbo, T.; Sato, M. *Synthesis* **1981**, 871.
6. Bedenbaugh, A. O; Bedenbaugh, J. H.; Bergin, W. A.; Adkins, J. D. *J. Am. Chem. Soc.* **1970**, *92*, 5774.
7. Brown, H. C.; Cha, J. S.; Nazer, B.; Yoon, N. M. *J. Am. Chem. Soc.* **1984**, *106*, 8001.
8. Fujisawa, T.; Mori, T.; Tsuge, S.; Sato, T. *Tetrahedron Lett.* **1983**, *24*, 1543.
9. Bohme, H.; Viehe, H. G. "Iminium Salts in Organic Chemistry;" Taylor, E. C., Ed.; In Advances in Organic Chemistry; Wiley: New York; 1976, Vol. 9, Part 1; 1979, Vol. 9, Part 2.

Appendix

Chemical Abstracts Nomenclature (Collective Index Number); (Registry Number)

6-Oxodecanal: Decanal, 6-oxo- (10); (63049-53-6)

6-Oxodecanoic acid: Decanoic acid, 6-oxo- (9); (4144-60-9)

Oxalyl chloride (8); Ethanedioyl dichloride (9); (79-37-8)

N,N-Dimethylchloromethylenammonium chloride: Ammonium, (chloromethylene)dimethyl-, chloride (8); Methanaminium, N-(chloromethylene)-N-methyl, chloride (9); (3724-43-4)

TRIPLE BOND ISOMERIZATIONS: 2- TO 9-DECYN-1-OL

$$HOCH_2C{\equiv}C(CH_2)_6CH_3 \xrightarrow[H_2N(CH_2)_3NH_2]{Li\ HN(CH_2)_3NH_2,\ KOtBu} HO(CH_2)_8C{\equiv}CH$$

Submitted by Suzanne R. Abrams and Angela C. Shaw.[1]

Checked by Maurizio Taddei and Ian Fleming.

1. Procedure

Caution: This preparation should be carried out in an efficient hood and the operator should wear gloves to protect against spillage of corrosive 1,3-diaminopropane.

A 1-L, three-necked, round-bottomed flask is equipped with a magnetic stirring bar, thermometer, pressure-equalizing dropping funnel to which an argon inlet is attached, and a condenser fitted with a drying tube filled with potassium hydroxide pellets (Note 1). The flask is charged with lithium (4.2 g, 0.6 mol, Note 2) and 1,3-diaminopropane (300 mL, Note 3). The mixture is stirred at room temperature for 30 min. A slight exothermic reaction takes place as the lithium dissolves.

The mixture is stirred and heated in an oil bath at 70°C until the blue color discharges (approximately 3 hr), affording a white suspension of the lithium amide. The reaction mixture is cooled to room temperature, and potassium tert-butoxide (44 g, 0.4 mol, Note 4) is added to the flask using a powder funnel. The resultant pale yellow solution is stirred for 20 min at room temperature, and then 2-decyn-1-ol (15.4 g, 0.1 mol, Note 5) is added over 10 min by means of the dropping funnel (Note 6). Residual 2-decyn-1-ol is washed into the flask with 1,3-diaminopropane (20 mL). The reddish brown mixture is stirred for 30 min and then poured into 1 L of ice water and extracted four times with 500-mL portions of hexane. The hexane extracts are combined and washed successively with 1 L each of water, 10% hydrochloric acid, and saturated sodium chloride solution. The hexane solution is dried over anhydrous sodium sulfate, filtered, and concentrated with a rotary evaporator. The crude product is distilled under reduced pressure to give 9-decyn-1-ol (12.8-13.5 g, 83-88%) (Notes 7, 8), as a colorless oil, bp 86-88°C/0.5 mm.

2. Notes

1. All glassware should be previously dried in an oven at 110°C for at least 2 hr.

2. Lithium wire (3.2 mm dia., 0.02% Na, Alfa Products, Morton/Thiokol Inc.) is cut into 1-cm pieces, then washed with hexane and quickly weighed into a tared beaker of hexane. The checkers also used lithium shot (BDH); the lithium did not all dissolve at room temperature, and heating at 70°C for 6 hr was necessary to discharge the blue color.

3. 1,3-Diaminopropane (97%, Aldrich Chemical Company, Inc.) was distilled at atmospheric pressure under nitrogen from barium oxide and stored over molecular sieves (4 Å).

4. Potassium tert-butoxide (Aldrich Chemical Company, Inc.) was used without further purification, and was added over 1-2 min.

5. 2-Decyn-1-ol (Farchan Labs., Lancaster Synthesis) was used without further purification. Upon addition of 2-decyn-1-ol to the reaction mixture slight warming is observed. The temperature is maintained at 25-30°C using a water bath.

6. The checkers used a syringe, injecting through a rubber septum.

7. The submitters used a Kugelrohr apparatus with an oven temperature of 80-90°C/0.05 mm.

8. The isomeric purity of the product is greater than 99% as determined by GLC of the trimethylsilyl ether. 9-Decyn-1-ol (n_D^{26} 1.4552) has the following spectroscopic characteristics: ^1H NMR ($CDCl_3$) δ: 1.1-1.7 (m, 12 H), 1.9 (t, J = 1.5, C≡CH), 2.1 (m, 2 H, $CH_2C≡C$), 3.6 (t, 2 H, J = 6, OCH_2); IR (film) ν_{max} cm^{-1}: 3400 (br), 3300 (s), 2100 (w), 1050 (s).

3. Discussion

In 1975 Brown and Yamashita [2] reported that a triple bond in any position of a straight chain hydrocarbon or acetylenic alcohol, when treated with a sufficiently strong base, could be isomerized exclusively to the free terminus of the chain. The "zipper reaction" thus provides a general solution to the problem of remote functionalization of a long hydrocarbon chain. Isomerizations along chains of thirty carbon atoms have been achieved. Isomerization is blocked by alkyl or hydroxyl branches; the triple bond then migrates to the free terminus.

The base employed by Brown and Yamashita was the potassium salt of 1,3-diaminopropane, prepared by reaction of potassium hydride with the solvent of the reaction, 1,3-diaminopropane. The reagent is very effective, and yields of isomerically pure products are high, but potassium hydride is hazardous, expensive and difficult to handle.

We and others have developed alternative methods for preparing the isomerization reagent.[3-6] Hommes and Brandsma[3] first made potassium or sodium amide in liquid ammonia and then replaced the solvent with 1,3-diaminopropane.

Kimmel and Becker[4] treated molten potassium or sodium with 1,3-diaminopropane in a flask immersed in an ultrasonic bath at 90°C. We[5,6] found that sodium hydride reacts with 1,3-diaminopropane, or 1,2-diaminoethane, on warming. All of these methods produce effective isomerization reagents with comparable conversions and yields. The lithium salt alone is a poor isomerization reagent; however, addition of potassium tert-butoxide affords a reagent which very effectively isomerizes triple bonds. This procedure, presented here, is our most refined method.[7] It is straightforward, gives reproducibly high yields, employs inexpensive reagents, and can be safely carried out on a large scale.

It is not satisfactory to employ 1,2-diaminoethane in place of 1,3-diaminopropane. The reagent is not as stable; addition of potassium tert-butoxide results in the immediate formation of a deep purple solution. Isomerizations proceed to completion, but yields are somewhat reduced (by about 10% in the case of the rearrangement of 2- to 9-decyn-1-ol).

1. Plant Biotechnology Institute, National Research Council of Canada, 110 Gymnasium Road, Saskatoon, Saskatchewan, S7N 0W9, Canada.

2. Brown, C. A.; Yamashita, A. *J. Am. Chem. Soc.* **1975**, *97*, 891-892.
3. Hommes, H.; Brandsma, L. *Recl. Trav. Chim. Pays-Bas* **1977**, *96*, 160.
4. Kimmel, T.; Becker, D. *J. Org. Chem.* **1984**, *49*, 2494-2496.
5. Macaulay, S. R. *J. Org. Chem.* **1980**, *45*, 734-735.
6. Abrams, S. R.; Nucciarone, D. D.; Steck, W. F. *Can. J. Chem.* **1983**, *61*, 1073-1076.
7. Abrams, S. R. *Can. J. Chem.* **1984**, *62*, 1333-1334.

Appendix

Chemical Abstracts Nomenclature (Collective Index Number); (Registry Number)

2-Decyn-1-ol (8,9); (4117-14-0)

9-Decyn-1-ol (8,9); (17643-36-6)

1,3-Diaminopropane: 1,3-Propanediamine (8,9); (109-76-2)

HOFMANN REARRANGEMENT UNDER MILDLY ACIDIC CONDITIONS USING [I,I-BIS(TRIFLUOROACETOXY)]IODOBENZENE: CYCLOBUTYLAMINE HYDROCHLORIDE FROM CYCLOBUTANECARBOXAMIDE

(Cyclobutanamine hydrochloride)

A. cyclobutane-COOH + EtOC(O)Cl + N-methylmorpholine ⟶ cyclobutane-C(O)OCO$_2$Et + N-methylmorpholinium Cl^-

cyclobutane-C(O)OCO$_2$Et + NH_3 ⟶ cyclobutane-C(O)NH$_2$ + CO_2 + C_2H_5OH

B. cyclobutane-C(O)NH$_2$ $\xrightarrow[2.\ H_2O]{1.\ C_6H_5I(O_2CCF_3)_2}$ cyclobutane-NH_3^+ Cl^- + CO_2 + C_6H_5I + CF_3CO_2H

Submitted by Merrick R. Almond, Julie B. Stimmel, E. Alan Thompson, and G. Marc Loudon.[1]

Checked by C. Eric Schwartz and Edwin Vedejs.

1. Procedure

A. Cyclobutanecarboxamide (Note 1). A 250-mL, round-bottomed flask, equipped with a mechanical stirrer and a drying tube (Drierite), is flame-dried and allowed to cool to room temperature. The flask is equipped with a punctured rubber septum, through which is inserted a -90°C thermometer. The flask is subjected to a nitrogen atmosphere by inserting a syringe needle connected to a nitrogen bubbler through the septum along with a second syringe needle used as an outlet. Under a flow of nitrogen the flask is charged via syringe with cyclobutanecarboxylic acid (6.0 g, 59.9 mmol, Note 2), 60 mL of dry tetrahydrofuran (Note 2), and N-methylmorpholine (6.6 mL, 59.9 mmol, Note 2). Stirring is commenced, and the solution is cooled to an internal temperature of -15°C using a dry ice-isopropyl alcohol bath at -20° to -25°C. Ethyl chloroformate (5.7 mL, 59.9 mmol, Notes 2, 3) is added and the solution is stirred for 5 min. The addition of ethyl chloroformate results in an internal temperature rise to +8° to +10°C and the precipitation of a white solid. Following the precipitation the continuously stirred mixture, still in the dry ice-isopropyl alcohol bath, is allowed to reach an internal temperature of -14°C. Anhydrous ammonia (Note 2), introduced into the flask via a syringe needle, is vigorously bubbled through the solution for 10 min with manual stirring; the internal temperature rises abruptly to 25°C. With the flask still in the cooling bath, stirring is continued for an additional 30 min, and the reaction mixture is stored in the freezer at -15°C overnight (Note 4).

The slurry is stirred with tetrahydrofuran (100 mL) at room temperature for 5 min and ammonium salts are removed by suction filtration through a Buchner funnel. After the solids are rinsed with tetrahydrofuran (20 mL), the filtrate is passed through a plug of silica gel (65 g Merck 60 230-400 mesh) in a coarse porosity sintered-glass filter funnel with aspirator suction. The funnel is further washed with acetonitrile (750 mL) and the combined filtrates are evaporated (rotary evaporator) to give a white solid. This material is recrystallized by heating (steam bath) with 8:1 ether:ethanol (70 mL); if necessary, ethanol is added dropwise to obtain a homogeneous solution. Cooling to room temperature results in white flakes which are collected by filtration (ether wash), 2.93 g. Two more crops (1.3 g total) are obtained by repeating the process for a total of 4.23 g (71%), mp 152-153°C (lit.[2] 155°C). In several similar runs, the yield of amide was 4.23-4.49 g (71-76%).

B. Cyclobutylamine hydrochloride. A 500-mL, round-bottomed flask is equipped with a magnetic stirring bar and covered with aluminum foil. To the flask is added a solution of [I,I-bis(trifluoroacetoxy)iodo]benzene (16.13 g, 37.5 mmol; Note 5) in 37.5 mL of acetonitrile, and the resulting solution is diluted with 37.5 mL of distilled deionized water. Cyclobutanecarboxamide (2.48 g, 25 mmol) is added; the amide quickly dissolves. Stirring is continued for 4 hr, and the acetonitrile is removed with a rotary evaporator. The aqueous layer is stirred with 250 mL of diethyl ether; to the stirring mixture is added 50 mL of concd hydrochloric acid (Note 6). The mixture is transferred to a separatory funnel and the layers are separated. The aqueous layer is extracted with two 125-mL portions of ether. The organic fractions are combined and extracted with 75 mL of 2N hydrochloric acid. The aqueous fractions are combined and concentrated with a rotary evaporator using a vacuum pump. Benzene (50 mL, Note 2) is added to the residue and the solution is concentrated with the rotary evaporator, again using a vacuum pump. Addition of benzene and concentration is repeated five more times. The crude solid is dried under reduced pressure over sulfuric acid overnight. To the product is added 5 mL of absolute ethanol and 35 mL of anhydrous ether, and the solution is heated at reflux on a steam bath. Ethanol is added slowly to the mixture, with swirling, until all the material is dissolved; the solution is cooled to room temperature. Anhydrous ether is added slowly until crystallization just begins. The flask is placed in the freezer and the product is allowed to crystallize. Filtration of the product and drying overnight under reduced pressure over phosphorus pentoxide yields 1.86-2.06 g of cyclobutylamine hydrochloride (69-77%), mp 183-185°C (lit.[3] 183-184°C).

2. Notes

1. The method used here for preparing the amide gives superior yields to two literature methods that employ the acid chloride as an intermediate.[3]

2. Sources and purification of reagents are as follows: Cyclobutanecarboxylic acid, 98%, is from Aldrich Chemical Company, Inc., and is vacuum distilled before use. Tetrahydrofuran is freshly distilled from sodium-benzophenone under nitrogen. N-Methylmorpholine, 99%, is from Aldrich Chemical Company, Inc., and is pre-dried over barium oxide, distilled from ninhydrin, and stored over sodium hydroxide pellets. Ethyl chloroformate, 97%, is from Aldrich Chemical Company, Inc., and is freshly distilled prior to use under a nitrogen atmosphere. Anhydrous ammonia (99.99% min) is from a Matheson lecture bottle. The silica gel used for flash chromatography is Davidson grade 62, 60-200 mesh. Acetonitrile is from Fisher Scientific Company, HPLC grade. Benzene, spectral grade, is from J. T. Baker.

3. The submitters have repeated this preparation with isobutyl chloroformate substituted for ethyl chloroformate with no increase in yield.

4. The submitters purified the product by flash chromatography as follows. Tetrahydrofuran is removed on a rotary evaporator. Silica gel (20 g) and 80 mL of acetonitrile (Note 2) are added to the flask. The resulting slurry is concentrated with a rotary evaporator to a dry solid. The material is scraped from the flask and loaded onto a flash chromatography column (50 mm

diameter) containing 250 g of silica gel, prepared according to the method of Still.[4] The column is eluted with acetonitrile (Note 2) at a flow rate of 0.5 in/min (Note 7). The first 500 mL of eluent is measured with a graduated cylinder and discarded, and then fractions (80 x 23 mL) are collected in test tubes. A small aliquot (5 μL) is taken from every other tube and these aliquots are spotted in successive lanes on a silica gel thin-layer chromatography plate (E. Merck No. 5735), which is developed with acetonitrile. The plate is dried and the product detected by the chlorine/starch-potassium iodide procedure (Note 8). This thin-layer analysis reveals that early fractions from the flash chromatography column contain a small amount of ethyl carbamate impurity at R_f = 0.53; another unidentified impurity (R_f = 0.37) follows. The product cyclobutanecarboxamide emerges next, beginning at about fraction 14 (R_f = 0.26). The product appears as a blue-black spot on a faint blue background. There is some overlap between the second impurity and the product. The fractions containing only product are pooled and concentrated in a 1-L, round-bottomed flask with a rotary evaporator to a white crystalline solid. This product is dried under reduced pressure overnight to yield cyclobutanecarboxamide (4.52 g, 76.1%), mp 156-157.5°C (lit.[2] 155°C).

5. [I,I-Bis(trifluoroacetoxy)iodo]benzene is prepared by dissolving, with heating, a given number of grams of (I,I-diacetoxyiodo)benzene (iodobenzene diacetate, Aldrich Chemical Company, Inc.; see also *Org. Synth., Collect. Vol. V*, **1973**, 660) in twice that number of milliliters of trifluoroacetic acid that has been distilled from a small amount of phosphorus pentoxide. For example, 40 g of (I,I,-diacetoxyiodo)benzene is dissolved in 80 mL of trifluoroacetic acid in an Erlenmeyer flask, which is allowed to stand in a dark drawer. The [I,I-bis(trifluoroacetoxy)iodo]benzene

crystallizes and is isolated by suction filtration within 2 hr (53-70% yield). If crystallization does not occur it can be induced by scratching or seeding. It has been the submitters' experience that if a dark yellow trifluoroacetic acid supernatant is obtained, the yield of the rearrangement reaction carried out with the resulting reagent is invariably poor; the supernatant solution is normally very lightly colored. If the proportion of trifluoroacetic acid is reduced, a greater weight of crystals is obtained; however, this material gives considerably lower yields in the rearrangement. Upon standing, particularly in the light, [I,I-bis(trifluoroacetoxy)-iodo]benzene turns yellow; this reagent also gives poor yields in the rearrangement and yellow reaction mixtures; the reaction mixtures of satisfactory rearrangements are water-white. The reagent should be stored in a dark bottle under nitrogen or argon.

The submitters have also found that the rearrangement can be effected with (I,I-diacetoxyiodo)benzene or iodosobenzene and two equivalents of trifluoroacetic acid.

6. Hydrochloric acid not only provides the chloride counter-ion for the final product, but also effects the removal of any unreacted (I,I-bis(tri-fluoroacetoxy)iodo]benzene as the ether-soluble (I,I-dichloroiodo)benzene (iodobenzene dichloride).

7. A nitrogen regulator has to be set at 8 p.s.i. to achieve a pressure sufficient to maintain a flow rate of 0.5 in/min. A slower flow rate results in poor resolution.

8. The thin-layer plate is chlorinated for 1 min by placing it in a chamber with potassium chlorate to which a few drops of hydrochloric acid are added. The plate is dried in air for 10 min and then sprayed with aqueous 1% starch-1% potassium iodide solution.

3. Discussion

Hypervalent iodine reagents have been used recently for a variety of organic transformations,[5] including α-hydroxylation of ketones.[6] (I,I-Dicarboxyiodo)benzene derivatives have also found a variety of uses.[7] The use of (I,I-diacetoxyiodo)benzene for the conversion of amides to carbamates in alcohol solvents was studied by Smith and Baumgarten,[8] and the "acidic Hofmann rearrangement" utilized in this preparation, in which amides are converted directly into the corresponding amines in partially aqueous solution, was developed by Loudon, et al.[9] and applied to a variety of amides; the mechanism of the rearrangement has also been studied.[10] The rearrangement occurs with retention of stereochemical configuration at the migrating alkyl group,[8,9b,11] and the relative rates of rearrangement generally follow the migratory preferences observed in the Lossen rearrangement, Baeyer-Villiger reaction, and similar migrations to electron-deficient centers.

A particularly interesting application of this reagent is the preparation of "geminal amino amides," a novel class of compounds that have surprising stability in aqueous solution.[12] These derivatives have found application in

$$CH_3C(=O)NHCH(CH(CH_3)_2)C(=O)NH_2 \xrightarrow[\text{2. HCl}]{\text{1. } C_6H_5I(O_2CCF_3)_2} CH_3C(=O)NHCH(CH(CH_3)_2)\overset{+}{N}H_3\ Cl^-$$

the construction of retro-inverso-peptides,[13] in the design of a novel class of prodrugs,[14] and as intermediates in a carboxyl-terminal peptide degradation.[15]

One important restriction is that the reaction cannot be applied to amides in which the carboxamide group is directly attached to an aromatic ring; in these cases, rearrangement occurs, but the resulting aromatic amine is more rapidly oxidized than the starting amide by remaining reagent.[16]

Although iodine(III) reagents attack double bonds, the rearrangement of the amide group is, at least in some cases, more rapid than electrophilic attack on alkenes. Thus 3-cyclohexene-1-carboxamide rearranges smoothly to the corresponding amine as long as only one equivalent of [I,I-bis(trifluoroacetoxy)iodo]benzene is used.

The acidic nature of the reagent is important; the trifluoroacetic acid liberated in the reaction catalyzes hydrolysis of the intermediate isocyanate, and also ensures that the amine which is formed is protonated and cannot react with the isocyanate to give urea by-products. The reaction can be accelerated by addition of pyridine to an observed pH of about 3.5, and is retarded by added acid or trifluoroacetate ion.[9b,10] In the present procedure pyridine was not employed, since the reaction in its absence proceeds with a satisfactory rate.

1. Department of Medicinal Chemistry and Pharmacognosy, School of Pharmacy and Pharmacal Sciences, Purdue University, West Lafayette, IN 47907.

2. Normant, H.; Voreux, G. *Compt. Rend.* **1950**, *231*, 703-704.
3. (a) Hanack, M.; Ensslin, H. M. *Justus Liebigs Ann. Chem.* **1966**, *697*, 100-110; (b) Dem'yanov, N. Ya.; Shuikina, Z. I. *J. Gen. Chem. (U.S.S.R.)* **1935**, *5*, 1213-1225; *Chem. Abstr.* **1936**, *30*, 1032[6].
4. Still, W. C.; Kahn, M.; Mitra, A. *J. Org. Chem.* **1978**, *43*, 2923-2925.
5. Varvoglis, A. *Chem. Soc. Rev.* **1981**, *10*, 377-407.

6. (a) Moriarty, R. M.; Hou, K.-C.; Prakash, I.; Arora, S. K. *Org. Synth.* **1985**, *64*, 138-145; (b) Moriarty, R. M.; Hu, H. *Tetrahedron Lett.* **1981**, *22*, 2747-2750.

7. Varvoglis, A. *Synthesis* **1984**, 709-726.

8. Smith, H. L. Ph. D. Disseration, University of Nebraska, 1970, pp. 64ff.

9. (a) Radhakrishna, A. S.; Parham, M. E.; Riggs, R. M.; Loudon, G. M. *J. Org. Chem.* **1979**, *44*, 1746-1747; (b) Loudon, G. M.; Radhakrishna, A. S.; Almond, M. R.; Blodgett, J. K.; Boutin, R. H. *J. Org. Chem.* **1984**, *49*, 4272-4276.

10. Boutin, R. H.; Loudon, G. M. *J. Org. Chem.* **1984**, *49*, 4277-4284.

11. Pallai, P. V.; Goodman, M. *J. Chem. Soc., Chem. Commun.* **1982**, 280-281.

12. (a) Bergmann, M.; Zervas, L.; Schneider, F. *J. Biol. Chem.* **1936**, 113, 341-357; (b) Loudon, G. M.; Jacob, J. *J. Chem. Soc., Chem. Commun.* **1980**, 377-378; (c) Loudon, G. M.; Almond, M. R.; Jacob, J. N. *J. Am. Chem. Soc.* **1981**, *103*, 4508-4515.

13. (a) Goodman, M.; Chorev, M. *Acc. Chem. Res.* **1979**, *12*, 71; (b) Chorev, M.; Wilson, C. G.; Goodman, M. *J. Am. Chem. Soc.* **1977**, *99*, 8075-8076.

14. (a) Bundgaard, H.; Johansen, M. *J. Pharm. Sci.* **1980**, *69*, 44; (b) Johansen, M.; Bundgaard, H. *Arch. Pharm. Chemi., Sci. Ed.* **1980**, *8*, 141.

15. (a) Loudon, G. M.; Parham, M. E. *Tetrahedron Lett.* **1978**, 437-440; (b) Parham, M. E.; Loudon, G. M. *Biochem. Biophys. Res. Commun.* **1978**, *80*, 1-6.

16. Barlin, G. B.; Pausacker, K. H.; Riggs, N. V. *J. Chem. Soc.* **1954**, 3122-3215.

Appendix

Chemical Abstracts Nomenclature (Collective Index Number); (Registry Number)

Cyclobutylamine hydrochloride (8); Cyclobutanamine hydrochloride (9); (6291-01-6)

[I,I-Bis(trifluoroacetoxy)iodo]benzene: Iodine, phenylbis(trifluoroacetato-O-)- (9); (2712-78-9)

Cyclobutanecarboxamide (8,9); (1503-98-6)

Cyclobutanecarboxylic acid (8,9); (3721-95-7)

Ethyl chloroformate: Formic acid, chloro-, ethyl ester (8); Carbonochloridic acid, ethyl ester (9); (541-41-3)

PREPARATION AND INVERSE ELECTRON DEMAND DIELS-ALDER REACTION OF AN ELECTRON-DEFICIENT HETEROCYCLIC AZADIENE: TRIETHYL 1,2,4-TRIAZINE-3,5,6-TRICARBOXYLATE

(1,2,4-Triazine-3,5,6-tricarboxylic acid, triethyl ester)

A. $N{\equiv}C\text{-}CO_2Et$ $\xrightarrow[H_2S]{Et_2NH}$ $H_2N\text{-}C(=S)\text{-}CO_2Et$

B. $H_2N\text{-}C(=S)\text{-}CO_2Et$ $\xrightarrow{N_2H_4}$ $H_2N\text{-}C(=N\text{-}NH_2)\text{-}CO_2Et$

C. $NaO_2C\text{-}C(OH)_2\text{-}C(OH)_2\text{-}CO_2Na$ $\xrightarrow[C_2H_5OH,\ 0^{\circ}C]{HCl(g)}$ $EtO_2C\text{-}CO\text{-}CO\text{-}CO_2Et$

D. $EtO_2C\text{-}CO\text{-}CO\text{-}CO_2Et$ + $H_2N\text{-}C(=N\text{-}NH_2)\text{-}CO_2Et$ $\xrightarrow{25^{\circ}C}$ triethyl 1,2,4-triazine-3,5,6-tricarboxylate

E. triethyl 1,2,4-triazine-3,5,6-tricarboxylate + N-vinyl-2-pyrrolidinone $\longrightarrow$ triethyl pyridine-2,3,6-tricarboxylate

Submitted by Dale L. Boger, James S. Panek, and Masami Yasuda.[1]

Checked by Pauline J. Sanfilippo and Andrew S. Kende.

1. Procedure[2a]

Caution! Hydrogen sulfide is highly toxic and a stench. Steps A and B must be run in an efficient fune hood.

A. Ethyl thioamidooxalate.[3] A 100-mL, round-bottomed flask is fitted with a magnetic stirring bar. Ethyl cyanoformate (20 g, 0.20 mol, Note 1) in benzene (25 mL) is added to the reaction vessel and the mixture is cooled to 0°C with an ice bath. Diethylamine (Note 2, 0.4 g, 5.5 mmol, 0.57 mL) is added to the stirring reaction mixture (0°C) and hydrogen sulfide (Note 3) is then bubbled into the reaction for an additional 15-20 min. The reaction mixture is allowed to stir at 25°C (14-16 hr). The crude product is collected by filtration (Note 4) and washed with benzene (2 x 3 mL) to give 20.96 g (78%) of pure ethyl thioamidooxalate. The filtrate is concentrated under reduced pressure and the crude product subjected to chromatography on silica gel (30% ether-hexane eluant) to give an additional 1.57 g of ethyl thioamidooxalate. The total amount of ethyl thioamidooxalate isolated as a bright yellow solid is 22.53 g (84%); mp 63-66°C (Note 5).

B. Ethyl oxalamidrazonate. A 1-L, round-bottomed flask is equipped with a magnetic stirring bar and fitted with a 125-mL addition funnel. A solution of anhydrous hydrazine (4.8 g, 0.15 mol) in ethanol (75 mL) is added dropwise (10 min) to a stirred solution of ethyl thioamidooxalate (20.0 g, 0.15 mol) in ethanol (450 mL) at 25°C. The reaction mixture is stirred at 25°C (3.0 hr). The solvent is removed under reduced pressure and the reddish-orange solid is triturated with ethanol (350 mL). The ethanolic solution containing the oxalamidrazonate is concentrated under reduced pressure to afford 13.90 g (71%) of ethyl oxalamidrazonate as a yellow solid (Note 6).

C. Diethyl dioxosuccinate. A 1-L, round-bottomed flask equipped with a magnetic stirring bar is charged with dihydroxytartaric acid disodium salt hydrate (100 g, 0.44 mol, Note 7) and absolute ethanol (750 mL, Note 8). The suspension is cooled to 0°C with an ice bath and anhydrous hydrogen chloride gas (Note 9) is bubbled into the reaction mixture with stirring (0°C, approximately 30 min). The reaction mixture is stoppered and placed in the refrigerator for 72 hr. The mixture is filtered using a Büchner funnel and the filtrate is concentrated under reduced pressure. The crude diethyl dioxosuccinate is distilled under reduced pressure to afford 39.60 g (44%) of pure diethyl dioxosuccinate (Note 10), bp 109-116°C (6-8 mm); lit.[4] bp 109-114°C (6 mm).

D. Triethyl 1,2,4-triazine-3,5,6-tricarboxylate. A 1-L, three-necked, round-bottomed flask is equipped with a magnetic stirring bar, 500-mL addition funnel and a nitrogen inlet. A solution of ethyl oxalamidrazonate (11.6 g, 88.0 mmol) in absolute ethanol (350 mL) is added dropwise (40-45 min) to a stirring solution of diethyl dioxosuccinate (23.1 g, 114.0 mmol) in absolute ethanol (86 mL) at 25°C under nitrogen. After the addition is complete the reaction mixture is stirred at 25°C (16 hr). A reflux condenser is fitted onto the three-necked, round-bottomed flask and the reaction mixture is warmed at reflux for 2.0 hr. The reaction mixture is cooled and the solvent is removed under reduced pressure. Purification of the product is effected by gravity chromatography (Note 11) on a 5.20 x 40.0-cm column of silica gel (10-40% ether-hexane gradient elution), collecting 100-mL fractions. The fractions are analyzed by thin-layer chromatography on silica gel (40% ether-hexane eluant). The fractions containing product are combined and the solvent is removed under reduced pressure to afford 14.70 g (56%) of pure triethyl 1,2,4-triazine-3,5,6-tricarboxylate as a viscous, yellow oil[2] (Note 12).

E. 2,3,6-Tricarboethoxypyridine. A 50-mL, round-bottomed flask is fitted with a magnetic stirring bar and a reflux condenser. Triethyl 1,2,4-triazine-3,5,6-tricarboxylate (1.49 g, 5.0 mmol) and chloroform (22.7 mL, Note 13) are added to the reaction vessel. N-Vinyl-2-pyrrolidone (2.22 g, 20 mmol, 2.3 mL, Note 14) is added to the solution and the reaction mixture is warmed at 60°C under an atmosphere of nitrogen for 26 hr. The solvent is removed under reduced pressure and the crude product subjected to gravity chromatography (Note 11) on a 2.7 x 32-cm column of silica gel (40-50% ether-hexane gradient elution), collecting 50-mL fractions. The fractions are analyzed by thin-layer chromatography on silica gel (50% ether-hexane eluant). The fractions containing product are combined and the solvent is removed under reduced pressure to afford 1.01-1.35 g (68-92%) of 2,3,6-tricarboethoxypyridine as a yellow oil (Note 15).

2. Notes

1. The submitters employed, without purification, ethyl cyanoformate purchased from Aldrich Chemical Company, Inc.

2. The submitters employed, without purification, diethylamine purchased from Aldrich Chemical Company, Inc.

3. Hydrogen sulfide gas was purchased from Burnox, Kansas City, MO. This reaction should be run in a fume hood.

4. In some instances, it is necessary to cool the flask (ice bath) containing the ethyl thioamidooxalate to promote crystallization of the product.

5. The product has the following spectral properties: ^1H NMR ($CDCl_3$) δ: 1.39 (t, 3 H, J = 8, CH_3), 4.33 (q, 2 H, J = 8, CH_2), 7.30-8.30 (br s, 2 H, NH_2), mp 63-66°C, lit.[2] mp 64-65°C.

6. *Caution: This reaction should be carried out in a fume hood. Ethyl oxalamidrazonate cannot be stored in solution for prolonged periods of time.*

7. The submitters employed dihydroxytartaric acid disodium salt hydrate purchased from Aldrich Chemical Company, Inc.

8. Ethanol was dried by distillation from magnesium turnings immediately before use.

9. Anhydrous hydrogen chloride gas purchased from Burnox, Kansas City, MO. was employed.

10. The ^1H NMR spectrum of this compound is as follows: ^1H NMR ($CDCl_3$) δ: 1.36 (t, 3 H, J = 8, CH_3), 4.44 (q, 2 H, J = 8, CH_2).

11. The checkers used flash chromatography for these steps.

12. The spectral properties of this product (orange oil) are as follows: ^1H NMR ($CDCl_3$) δ: 1.45 (t, 3 H, J = 7, CH_3), 1.48 (t, 3 H, J = 7, CH_3), 1.51 (t, 3 H, J = 7, CH_3); 4.38-4.68 (3 overlapping q, 6 H, three CH_2); IR (film) ν_{max} cm^{-1}: 2986, 1757, 1738, 1518, 1468, 1408, 1383, 1302, 1217, 1177, 1155, 1099, 1017, 857.

13. The submitters employed chloroform obtained from Fisher Chemical Co.

14. The submitters employed, without purification, N-vinyl-2-pyrrolidone obtained from GAF Corporation.

15. The spectral properties of the product are as follows: ^1H NMR ($CDCl_3$) δ: 1.38 (t, 3 H, J = 7, CH_3), 1.41 (t, 3 H, J = 7, CH_3), 1.43 (t, 3 H, J = 7, CH_3), 4.38 (q, 2 H, J = 7, CH_2), 4.44 (q, 2 H, J = 7, CH_2), 4.47 (q, 2 H, J = 7, CH_2), 8.16 (d, 1 H, J = 8, aromatic), 8.30 (d, 1 H, J = 8, aromatic); IR (film) ν_{max} cm^{-1}: 2986, 1728, 1586, 1570, 1468, 1455, 1406, 1387, 1370, 1321, 1283, 1239, 1152, 1071, 1021, 853, 762.

3. Discussion

This procedure describes the preparation of an electron-deficient heterocyclic azadiene suitable for use in inverse electron demand ($LUMO_{diene}$ controlled)[5] Diels-Alder reactions with electron-rich dienophiles.

Table I[6,7] details representative examples of the [4 + 2] cycloaddition of triethyl 1,2,4-triazine-3,5,6-tricarboxylate with pyrrolidine enamines and related electron-rich olefins. Cycloaddition occurs across carbon-3 and carbon-6 of the 1,2,4-triazine nucleus, and the nucleophilic carbon of the dienophile attaches to carbon-3 (eq 1). Loss of nitrogen from the initial adduct and aromatization with loss of pyrrolidine affords pyridine products.

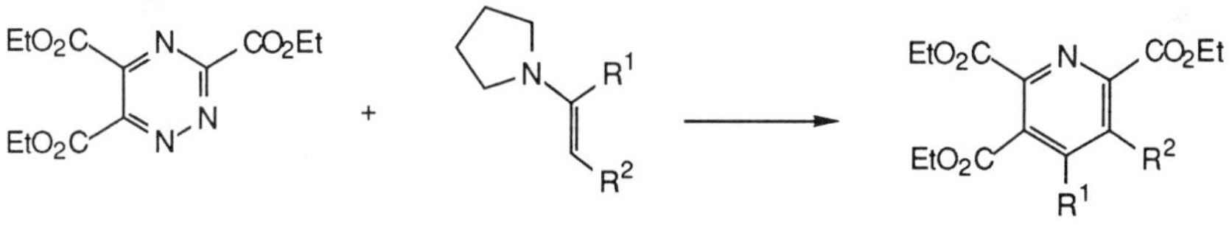

(1)

Similar reactivity and regioselectivity is observed with the parent system, 1,2,4-triazine (eq 2).[8a] Reduction of this process to a catalytic Diels-Alder reaction with in situ generation of the pyrrolidine enamine does not alter these observations (eq 3).[8b]

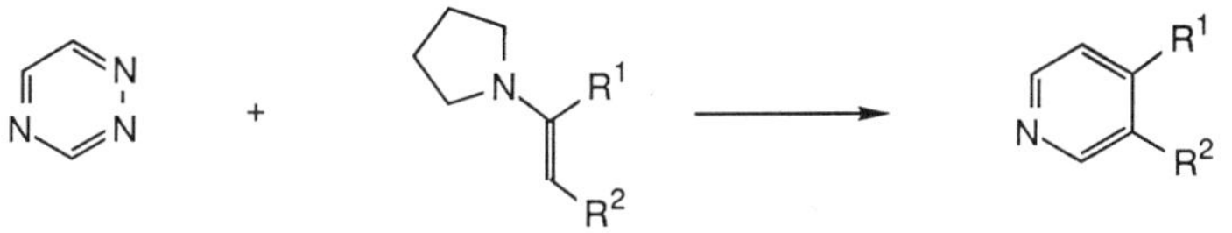

(2)

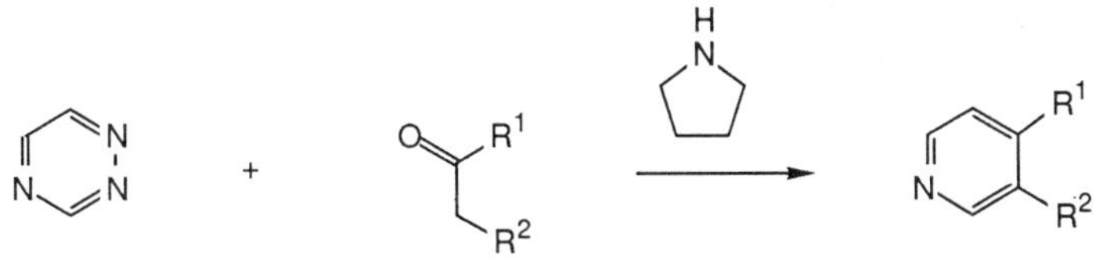

(3)

The number and position of electron-withdrawing substituents on the 1,2,4-triazine nucleus and the reactivity of the electron-rich dienophile determine the mode of cycloaddition (additions across C-5/N-2 as well as C-3/C-6 of the 1,2,4-triazine nucleus have been observed) as well as the regioselectivity.[8,9] A complete survey of the reported Diels-Alder reactions of 1,2,4-triazines including triethyl 1,2,4-triazine-3,5,6-tricarboxylate has been compiled.[10]

1. Department of Medicinal Chemistry, University of Kansas, Lawrence, KS 66045. Present address: Department of Chemistry, Purdue University, West Lafayette, IN 47907.
2. The procedure described is a modification of a detailed preparation: (a) Ratz, R.; Schroeder, H. *J. Org. Chem.* **1958**, *23*, 1931; (b) For the preparation of 5,6-dimethoxycarbonyl-3-ethoxycarbonyl-1,2,4-triazine, see: Martin, J. C. *J. Org. Chem.* **1982**, *47*, 3761.
3. Boon, W. R. *J. Chem. Soc.* **1945**, 601.
4. Fox, H. H. *J. Org. Chem.* **1947**, *12*, 535.
5. Houk, K. N. *J. Am. Chem. Soc.* **1973**, *95*, 4092.
6. Boger, D. L.; Panek, J. S. *J. Org. Chem.* **1982**, *47*, 3763.
7. Boger, D. L.; Panek, J. S. *J. Am. Chem. Soc.* **1985**, *107*, 5745; Boger, D. L.; Panek, J. S. *J. Org. Chem.* **1983**, *48*, 621.
8. (a) Boger, D. L.; Panek, J. S. *J. Org. Chem.* **1981**, *46*, 2179; (b) Boger, D. L.; Panek, J. S.; Meier, M. M. *J. Org. Chem.* **1982**, *47*, 895; (c) Boger, D. L.; Duff, S. R.; Panek, J. S.; Yasuda, M. *J. Org. Chem.* **1985**, *50*, 5782, 5790

9. Neunhoeffer, H.; Wiley, P. F. "Chemistry of 1,2,3-Triazines and 1,2,4-Triazines, Tetrazines and Pentazines", Wiley: New York, 1978, p. 189.

10. Boger, D. L. *Tetrahedron*, **1983**, *39*, 2869.

Appendix

Chemical Abstracts Nomenclature (Collective Index Number); (Registry Number)

Triethyl 1,2,4-triazine-3,5,6-tricarboxylate: 1,2,4-Triazine-3,5,6-tricarboxylic acid, triethyl ester (10); (74476-38-3)

Ethyl thioamidooxalate: Oxamic acid, 2-thio-, ethyl ester (8); Acetic acid, aminothioxo-, ethyl ester (9); (16982-21-1)

Ethyl cyanoformate: Formic acid, cyano-, ethyl ester (8); Carbonocyanidic acid, ethyl ester (9); (623-49-4)

Ethyl oxalamidrazonate: Acetic acid, hydrazinoimino-, ethyl ester (9); (53085-26-0)

Diethyl dioxosuccinate: Butanedioic acid, dioxo-, diethyl ester (9); (59743-08-7)

Dihydroxytartaric acid, disodium salt hydrate: Butanedioic acid, tetrahydroxy-, disodium salt (9); (866-17-1)

N-Vinyl-2-pyrrolidone: 2-Pyrrolidinone, 1-vinyl- (8); 2-Pyrrolidinone, 1-ethenyl- (9); (88-12-0)

TABLE I

DIELS-ALDER REACTION OF TRIETHYL 1,2,4-TRIAZINE-3,5,6-TRICARBOXYLATE

	Dienophile	Conditions: Solv., Temp., Time	Product	% Yield
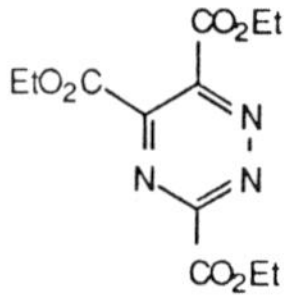 EtO$_2$C, CO$_2$Et, N, N, N, CO$_2$Et	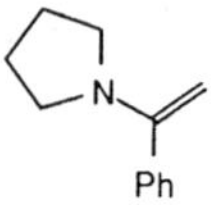 N, Ph	$CHCl_3$ 60°C 18 hr	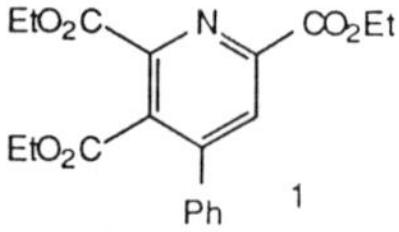 EtO$_2$C, N, CO$_2$Et, EtO$_2$C, Ph 1	79
	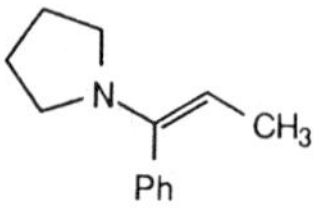 N, CH$_3$, Ph	$CHCl_3$ 45°C 8 hr	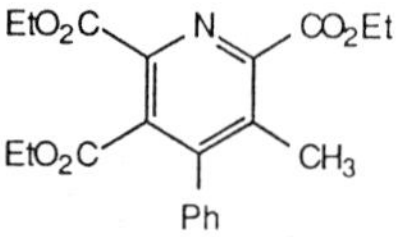 EtO$_2$C, N, CO$_2$Et, EtO$_2$C, CH$_3$, Ph 2	73
	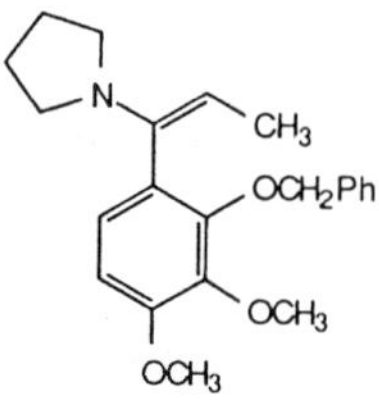 N, CH$_3$, OCH$_2$Ph, OCH$_3$, OCH$_3$	$CHCl_3$ 45°C 3 hr	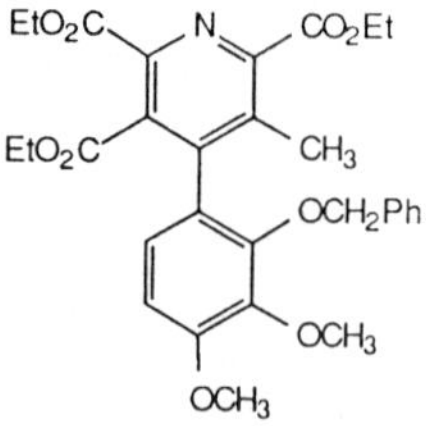 EtO$_2$C, N, CO$_2$Et, EtO$_2$C, CH$_3$, OCH$_2$Ph, OCH$_3$, OCH$_3$ 3	59
	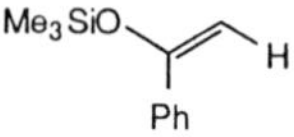 Me$_3$SiO, H, Ph	$CHCl_3$ 60°C 22 hr	1	84
	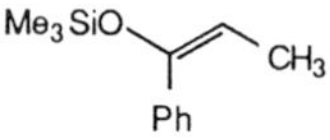 Me$_3$SiO, CH$_3$, Ph	$CHCl_3$ 60°C 16 hr	2	0
	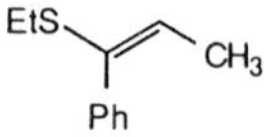 EtS, CH$_3$, Ph	$CHCl_3$ 80-160°C 10-20 hr	3	0
	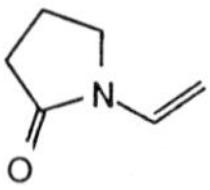 N, O	$CHCl_3$ 60°C 26 hr	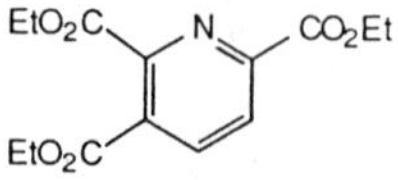 EtO$_2$C, N, CO$_2$Et, EtO$_2$C	92

(S)-2-CHLOROALKANOIC ACIDS OF HIGH ENANTIOMERIC PURITY FROM (S)-2-AMINO ACIDS: (S)-2-CHLOROPROPANOIC ACID

(Propanoic acid, 2-chloro-, (S)-)

$$H_2N-C(COOH)(R)-H \xrightarrow[HCl]{NaNO_2} Cl-C(COOH)(R)-H$$

R = $-CH_3$, $-CH(CH_3)_2$, $-CH_2CH(CH_3)_2$, (S)-$CH(CH_3)CH_2CH_3$

Submitted by Bernhard Koppenhoefer and Volker Schurig.[1]

Checked by G. Nagabhushana Reddy and James D. White.

1. Procedure

(S)-2-Chloropropanoic acid. In a 4-L, three-necked, round-bottomed flask equipped with a mechanical stirrer, 500-mL dropping funnel and a two-necked adapter fitted with a thermometer and reflux condenser (Note 1), 89.1 g (1 mol) of (S)-alanine (Note 2) is dissolved in 1300 mL of 5 N hydrochloric acid (Note 3). The mixture is cooled to 0°C in an ice/sodium chloride bath (Note 4) and a precooled solution of 110 g (1.6 mol) of sodium nitrite in 400 mL of water is added dropwise at a rate of about 2 mL/min under vigorous stirring and efficient cooling so that the temperature of the reaction mixture is kept below 5°C. After 5 hr, the bath is removed and the reaction is allowed to stand overnight at room temperature (Note 5). The reflux condenser is connected with a water aspirator and the flask is carefully evacuated with stirring for 3 hr to remove nitrogen oxides, whereupon the color changes from yellowish brown to pale yellow. While the mixture is stirred vigorously, 100 g of solid sodium carbonate is added carefully in small portions so as to prevent excessive foaming. The reaction mixture is extracted with four portions of 400 mL of diethyl ether. The combined ether layers are concentrated to ca. 300 mL using a rotary evaporator at atmospheric pressure. The solution is washed with 50 mL of saturated brine which thereafter is reextracted with three portions of 100 mL of diethyl ether. The combined ethereal solutions are dried for 10 hr over calcium chloride. The ether is distilled off with a rotary evaporator at atmospheric pressure (bath temperature 40 to 50°C). The oily residue is transferred into a distillation flask (rinsing the remainder with small portions of ether) and then fractionally distilled at reduced pressure, the main fraction boiling within a range of 2 to 3°C (i.e., bp 75-77°C at 10 mm) (Note 6) to give 63-71 g (58-65%) of an oil. The colorless oil is sufficiently pure for most purposes (Notes 7 and 8). On prolonged standing in a refrigerator, the oil tends to solidify partially or totally, but the white crystals formed have no sharp melting point. This procedure can be employed for other α-amino acids (see Table and the Discussion).

2. Notes

1. If the procedure is carried out under an atmosphere of nitrogen, oxidation of nitrogen monoxide to nitrogen dioxide is prevented and the reaction mixture remains colorless, but the yield is not improved.

2. The checkers used (S)-alanine of 97% optical purity, purchased from Aldrich Chemical Company, Inc.

The enantiomeric purities of the (S)-amino acids were checked by preparing the corresponding N-trifluoroacetylamino acid methyl esters, which are resolved into enantiomers by gas liquid chromatography on glass capillary columns coated with the chiral stationary phase 'Chirasil-Val',[2] (see Table). For this purpose, an aliquot of the aqueous solution, containing about 0.1 to 1 mg of the amino acid, is transferred to a 1-mL vial. Water is removed by a stream of nitrogen and the residue is transformed to the methyl ester hydrochloride (15% hydrochloric acid in methanol, 110°C, 30 min) and finally (after drying in a stream of nitrogen) to the N-trifluoroacetyl derivative (trifluoroacetic anhydride, 110°C, 10 min). This material is dried and dissolved in dichloromethane for GLPC analysis.

The commercially-available (S)-amino acids alanine, valine, leucine and isoleucine usually contained only negligible amounts (a few parts/thousand) of the (R)-antipode, but occasionally up to 2.5% of the (R)-enantiomer has been detected in (S)-alanine and (S)-valine. The (R)-enantiomer is almost completely removed by one recrystallization from water.

3. Concentrated hydrochloric acid is diluted by its own volume with water. Hydrochloric acid (2.4 L, 5 N) is employed for the less soluble (S)-amino acids valine, leucine and isoleucine.

4. A precipitate of the amino acid hydrochloride which formed on cooling is dissolved during the reaction.

5. The less soluble chloroalkanoic acids (R larger than methyl) separate from the solution as an oil.

6. Sometimes a brownish forerun is observed (bp up to 70°C/10 mm for 2-chloropropanoic acid), turning green in a refrigerator and occasionally undergoing vigorous decomposition. It is therefore recommended that distillation be interrupted and the flask containing the forerun removed.

7. If a brownish color or a wide range of the boiling point of the main fraction is observed, redistillation is recommended. Redistillation is necessary for (2S,3S)-2-chloro-3-methylpentanoic acid. Yields are given in Table. Enantiomeric purities (see Table) were determined after conversion to tert-butyl amides (catalyzed by dicyclohexylcarbodiimide, 30 min at 0°C in dichloromethane) by gas liquid chromatography on 'Chirasil-Val'.[3] The chiroptical data were determined on double-distilled (S)-2-chloroalkanoic acids. Traces of water (determined by gas liquid chromatography on Porapak using a thermal conductivity detector) cause a significant decrease of the specific rotation.

8. The spectral properties of (S)-2-chloropropanoic acid were as follows: ^{1}H NMR ($CDCl_3$) δ: 1.66 (d, 3 H, J = 6.7), 4.40 (q, 1 H, J = 6.7), 12.0 (s, 1 H; this signal may be broadened and shifted upfield due to minimal amounts of water); ^{13}C NMR ($CDCl_3$) δ: 20.9, 52.0, 176.0.

TABLE

(2S)-2-CHLOROALKANOIC ACIDS

```
   COOH
    |
Cl-C-H
    |
    R
```

Reactant	Substituent R	Yield (%)	bp (°C/mm)	e.e. (%) of Chloro Acid[a]	d_4^{20} (g/cm^{-3})	$[\alpha]_D^{20}$ (°)[e]
(S)-Alanine	$-CH_3$	64 ± 6	75-77/10	95.6[b]	1.265	-13.98
(S)-Valine	$-CH(CH_3)_2$	62 ± 5	103-105/10	97.7[b]	1.140	- 1.44
(S)-Leucine	$-CH_2CH(CH_3)_2$	58 ± 4	113-115/10	95.8[c]	1.082	-31.73
(2S)-Isoleucine	(3S)-$CH(CH_3)CH_2CH_3$	59 ± 6	111-112/10	98.3[c,d]	1.115	- 4.78

[a]In each case the starting amino acid was ≥ 99.8% optically pure, as shown by gas chromatography of the trifluoroacetyl methyl esters on Chirasil-Val (Note 2).

[b]By gas chromatography of the tert-butyl amides on Chirasil-Val (Note 7).

[c]By gas chromatography of the tert-butyl amides on (R)-N-lauroyl-1-(1-naphthyl)ethylamine (ref 8).

[d]Diastereomeric excess, referring to (2R,3S)-2-chloro-3-methylpentanoic acid. Total composition: 99.0% 2S,3S; 0.8% 2R,3S; 0.2% 2S,3R; approximately 0% 2R,3R. The starting amino acid was contaminated with 0.2% of the 2S,3R.

[e]Rotations were measured on the neat liquids; specific rotations are given for material of the indicated e.e.

3. Discussion

The present procedure is based on the method published by Fu, Birnbaum and Greenstein.[4] The yields are increased by the very slow addition of an aqueous solution of sodium nitrite to the reaction mixture as well as by a modified work-up procedure, i.e., careful removal of nitrogen oxides and the final decomposition of their adducts with carboxylic acids by buffering with sodium carbonate.

By using high-efficiency capillary gas chromatography with chiral stationary phases (i.e., 'Chirasil-Val'[2] and (R)-N-lauroyl-1-(1-naphthyl)-ethylamine[8]), it has been possible for the first time to determine the degree of racemization during the substitution reaction which proceeds with overall retention of configuration because of double inversion via an unstable α-lactone.[10] Thus, the maximum degree of inversion amounts to approximately 2.2%, resulting in a 2-chloroalkanoic acid of e.e. 95.6% (see Table) if racemization occurs in the diazotization reaction, and not in the conversion of the free chloroalkanoic acid to the tert-butylamide employed for analysis of the enantiomeric composition.[3,8] The enantiomeric yields given in the Table represent the *lowest* values found in various experiments. The degree of racemization at carbon atom 2 is strongly affected by the alkyl group R (see Table). Racemization is more pronounced in the case of less hindered primary and secondary carbon atoms adjacent to the stereocenter. It is interesting to note that the degree of racemization observed in the diazotization reaction runs parallel to the degree of racemization observed in aqueous solutions of the amino acids at pH 7.6 at elevated temperature.[11]

The diazotization in 5 N hydrochloric acid is superior to that in aqua regia[6] where up to 10% inversion has been observed.[12]

The method described may also be used for the preparation of the corresponding (R)-2-chloroalkanoic acids when starting from unnatural (R)-2-amino acids. For instance, (R)-2-aminodecanoic acid has been obtained in high enantiomeric yield by enzymatic cleavage of the racemic N-chloroacetyl derivative.[13] For amino acids containing large alkyl side chains diazotization at higher dilution is recommended. For the synthesis of racemic 2-chloroalkanoic acids the diazotization method described here appears more convenient than the direct chlorination of alkanoic acids.[14]

2-Chloroalkanoic acids bearing chiral side groups are useful starting materials for the synthesis of chiral alcohols of high enantiomeric purity. Thus, (3S)-3-methylpentanol-1 has been obtained from (2S,3S)-isoleucine via exhaustive lithium aluminum hydride reduction of the chloro acid.[15] Similarly, (3S)-1,3-butanediol has been obtained from (2S,3S)-allothreonine.[16] The time-controlled lithium aluminum hydride reduction of 2-chloroalkanoic acids leads to 2-chloro-1-alkanols (chlorohydrins) which can be cyclized to alkyloxiranes of high enantiomeric purity.[17]

1. Institut für Organische Chemie der Universität, Auf der Morgenstelle 18, D-7400 Tübingen, Federal Republic of Germany. We thank Mr. E. Koch and Professor E. Bayer, University of Tübingen, FRG, and Dr. K. Watabe and Professor E. Gil-Av, Weizmann Institute of Science, Rehovot, Israel, for the determination of the enantiomeric purity of the 2-chloroalkanoic acids. We thank Deutsche Forschungsgemeinschaft and Fonds der chemischen Industrie for support of this work.
2. Frank, H.; Nicholson, G. J.; Bayer, E. *J. Chromatogr. Sci.* **1977**, *15*, 174.
3. Koch, E.; Nicholson, G. J.; Bayer, E. *J. High. Resolut. Chromatogr. Chromatogr. Commun.* **1984**, *7*, 398.
4. Fu, S.-C. J.; Birnbaum, S. M.; Greenstein, J. P. *J. Am. Chem. Soc.* **1954**, *76*, 6054.
5. Klebe, J. F.; Finkbeiner, H. *J. Am. Chem. Soc.* **1968**, *90*, 7255.
6. Karrer, P.; Reschofsky, H.; Kaase, W. *Helv. Chim. Acta* **1947**, *30*, 271.
7. Gaffield, W.; Galetto, W. G. *Tetrahedron* **1971**, *27*, 915.
8. Koppenhoefer, B.; Watabe, K.; Gil-Av, E., submitted for publication.
9. Horeau, A. *Tetrahedron Lett.* **1969**, 3121.
10. Brewster, P.; Hiron, F.; Hughes, E. D.; Ingold, C. K.; Rao, P. A. D. S. *Nature* **1950**, *166*, 179.
11. Smith, G. G.; Sivakua, T. *J. Org. Chem.* **1983**, *48*, 627.
12. Schurig, V.; Koppenhoefer, B.; Buerkle, W. *Angew. Chem., Intern. Ed. Engl.* **1978**, *17*, 937.
13. Masaoka, Y.; Sakakibara, M.; Mori, K. *Agric. Biol. Chem.* **1982**, *46*, 2319.
14. Ogata, Y.; Sugimoto, T.; Inaishi, M. *Org. Synth.* **1980**, *59*, 20.
15. (a) Koppenhoefer, B.; Weber, R.; Schurig, V. *Synthesis* **1982**, 316; (b) Schurig, V.; Leyrer, U.; Wistuba, D. *J. Org. Chem.* **1986**, *51*, 242.

16. Hintzer, K.; Koppenhoefer, B.; Schurig, V. *J. Org. Chem.* **1982**, *47*, 3850.

17. Following procedure.

Appendix

Chemical Abstracts Nomenclature (Collective Index Number); (Registry Number)

(S)-2-Chloropropanoic acid: Propionic acid, 2-chloro, (S)- (8); Propanoic acid, 2-chloro-, (S)- (9); (29617-66-1)

(S)-Alanine: L-Alanine (8,9); (56-41-7)

(R)-ALKYLOXIRANES OF HIGH ENANTIOMERIC PURITY FROM (S)-2-CHLOROALKANOIC ACIDS VIA (S)-2-CHLORO-1-ALKANOLS: (R)-METHYLOXIRANE

(Oxirane, methyl- (R)-)

$$Cl-C(COOH)(R)-H \xrightarrow[Et_2O]{LiAlH_4} Cl-C(CH_2OH)(R)-H \xrightarrow[H_2O]{KOH} (R)\text{-}H_2C\text{-}O\text{-}CH(R)$$

$R = -CH_3, -CH(CH_3)_2, -CH_2CH(CH_3)_2, (S)-CH(CH_3)CH_2CH_3$

Submitted by Bernhard Koppenhoefer and Volker Schurig.[1]

Checked by G. Nagabhushana Reddy and James D. White.

1. Procedure

CAUTION: Methyloxirane is a suspected carcinogen for humans.

(S)-2-Chloropropan-1-ol. Into a 2-L, three-necked, round-bottomed flask equipped with a mechanical stirrer, 250 mL dropping funnel, stopper (Note 1) and an efficient reflux condenser fitted with a calcium chloride drying tube, is placed 9.1 g (0.24 mol) of lithium aluminum hydride; 400 mL of dry diethyl ether is added with caution. The slurry is cooled in an ice bath and a solution of 21.7 g (0.20 mol) of (S)-2-chloropropanoic acid (Note 2) in 150 mL of dry diethyl ether is added carefully with vigorous stirring over a 10-min period so that refluxing of the solvent is kept under control. After a total reaction time of 15 min (Note 3), the drying tube is removed and 20 mL of water is added drop by drop (*Caution: vigorous evolution of hydrogen!*) with efficient stirring and cooling (Note 4). The precipitate is dissolved (Note 5) by addition of 0.6 L of 2 N sulfuric acid (Note 6). The layers are

separated, and the aqueous layer is extracted with two 200-mL portions of diethyl ether. The combined ether layers are washed with 50 mL of water, 50 mL of sodium carbonate solution (Note 4) and 50 mL of sodium bicarbonate solution, each aqueous layer being reextracted with two 50-mL portions of diethyl ether (Note 7). The combined ethereal layers are concentrated with a rotary evaporator at atmospheric pressure (bath temperature 40 to 50°C) to approximately 300 mL, dried over sodium sulfate, and concentrated to give an oily residue. Fractional distillation at reduced pressure (Note 8) affords 10.6-11.0 g (56-58%) of a colorless oil. This procedure can be applied to chlorohydrins with other alkyl residues (see Notes 3 and 8, Table I).

(R)-Methyloxirane (Note 9). The reaction is conveniently carried out in a special apparatus (see Figure 1) in order to prevent loss of the volatile oxirane. A 50-mL, narrow-necked vessel (A) is equipped with a magnetic stirrer and a small Claisen stillhead (B), fitted with a thermometer and connected to a small receiver adapter with vacuum connection (C). A 25-mL or 50-mL flask (D) serves as a trap for the oxirane. To prevent clogging of the inlet pipe (E) by solidified reaction product, an appropriate flask (D) is chosen so that the distance between the inlet pipe (E) and the flask (D) is approximately 5 to 10 mm. The vacuum end of the adapter (C) is connected via a stopcock (F), a T-piece carrying a needle valve (G), and a manometer (H) to a water aspirator (I). The reaction vessel (A) is equipped with an ice bath, thermometer and a combined heater and magnetic stirrer which is placed on a jack. After the entire apparatus is connected (F closed), trap (D) is air-cooled in a Dewar (K) which is partially filled with liquid nitrogen, and heat-insulated at the top with cotton. A low-temperature thermometer (L) is placed at the same height near trap (D). The temperature of trap (D) is controlled, by moving the jack to the appropriate height to approximately

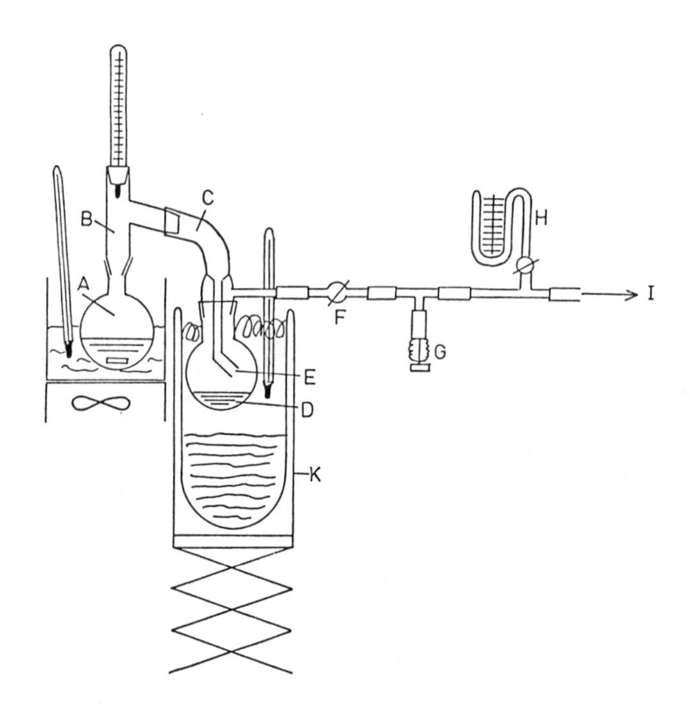

Figure 1. Apparatus for the Preparation of (R)-Methyloxirane.

-80°C. The pressure is adjusted by a needle valve (G) to 100 mm (F remains closed). A solution of 12.3 g (0.22 mol) of potassium hydroxide pellets in 12 mL of water is placed in vessel (A) and cooled to 0°C. Neat (S)-2-chloropropan-1-ol, 11.8 g (0.125 mol), is poured at once into the alkaline solution (Note 10), and the reaction vessel is immediately fitted with still-head (B) and stirred vigorously with efficient cooling. Stopcock (F) is opened occasionally for a short period until the pressure in the closed system is reduced to 100 mm (Note 11). The ice bath is replaced by a water bath at 20°C. As the cyclization reaction proceeds a white precipitate of potassium chloride is formed. After 10 min, the temperature of the bath is raised slowly to 30°C. Gentle boiling of the oxirane is maintained by cautiously opening stopcock (F) from time to time, attention being paid to the reaction vessel. After a total reaction time of 40 min (Note 12), air is allowed to enter the closed system at the top of the stillhead, and trap (D) is allowed to warm (Note 13) until two liquid phases are formed. The lower phase containing water is transferred via a Pasteur pipette into a small flask (Note 14). Flask (D) containing 5.9 g (81%) of crude (R)-methyloxirane is used in position (A) of the clean, dry apparatus (see Figure 1) for redistillation of the oxirane from calcium hydride. At atmospheric pressure (stopcock F open), flask (A) is cooled to 0°C, whereas trap (D) is kept at room temperature. Calcium hydride is added in small portions over a period of 1 to 2 hr until evolution of hydrogen ceases. Stopcock (F) is closed, trap (D) is cooled and the oxirane is distilled as described. Reduced pressure is applied with great care to avoid too vigorous boiling; 4.7-5.0 g (65-70%) of anhydrous oxirane is obtained as a clear liquid. This procedure can be employed for other oxiranes with slight modifications (see Notes 12 and 14, and Table II).

2. Notes

1. As a safeguard it is recommended that the reaction be performed under nitrogen, using a gas inlet instead of the stopper. The flask should be dry and free of faults.

2. (S)-2-Chloroalkanoic acids are prepared according to the procedure given previously.[2]

3. A total reaction time of 30 min is needed for more sterically hindered (S)-2-chloroalkanoic acids. Prolonged reaction time should be avoided to prevent hydrogenolysis of the chlorine-carbon bond.

4. Prolonged exposure to alkaline conditions should be avoided to prevent oxirane formation at this step.

5. The aqueous phase is allowed to remain opalescent, to avoid unnecessarily low pH-values.

6. Concentrated sulfuric acid, 60 g, is added to a beaker charged with 540 g of crushed ice. Precooled 2 N sulfuric acid is added to the reaction mixture.

7. Less than 5% of the chloroalkanoic acid is reisolated after acidification of the sodium carbonate phase and extraction with diethyl ether.

8. (S)-2-Chloropropan-1-ol is carefully distilled at atmospheric pressure using a 20-cm Vigreux column. (S)-2-Chloro-3-methylbutan-1-ol, (S)-2-chloro-4-methylpentan-1-ol and (2S,3S)-2-chloro-3-methylpentan-1-ol are distilled under reduced pressure with a spinning-band columm or a 'Spaltrohr-column' (approximately 50 theoretical plates, supplier: W. G. Fischer, D-5309 Meckenheim, FRG), see Table I. The main fractions are > 99% pure by GLC (OV 17 on Chromosorb P AW-DMCS). Because of the low boiling points of the oxiranes, diethyl ether should be completely removed from the chlorohydrins.

9. The synthesis should be carried out in a well-ventilated hood. *CAUTION: Methyloxirane is a suspected carcinogen for humans.*

10. The amount of chlorohydrin used is determined from the weight remaining in the original flask (approximately 1 g).

11. During the course of the reaction, stopcock (F) should remain closed except for short periods in order to avoid loss of the volatile oxirane. There is no danger of excess pressure in the closed system as long as trap (D) is cooled efficiently.

12. The chlorohydrins show different rates of cyclization, reflecting the steric hindrance of residue R. The most vigorous reaction is observed in the case of (S)-2-chloropropan-1-ol (R = CH_3); only 30°C at 100 mm is required. For (R)-isopropyloxirane, the temperature of the bath is raised slowly to 50°C, and after 40 min to 60°C, while the pressure is reduced to 50 mm for an additional 5 min. For the higher boiling oxiranes, e.g., (R)-isobutyloxirane and (S)-sec-butyl-(R)-oxirane [(2R,3S)-3-methyl-1,2-epoxy-butane], the temperature is raised slowly to 60°C within 1 hr, while the pressure is reduced carefully to 30 mm.

13. Build-up of pressure of methyloxirane is prevented by briefly opening the apparatus from time to time.

14. For methyloxirane, the binary system with water has been studied in detail.[3] By careful operation during the distillation, water is largely retained in the original flask (A). The racemate melts at -112°C, but the hydrate $C_3H_6O(H_2O)_{16}$ (mp -3°C) may solidify in the inlet tube (E). In the case of higher boiling oxiranes, substantial amounts of water are co-distilled. After removal from flask (D), the aqueous phase may be saturated with sodium chloride. Thereby a second portion of the oxirane (approximately 0.2 g) is separated and combined with the main portion in flask D.

3. Discussion

The method described here illustrates the transformation of optically active 2-chlorocarboxylic acids, which are readily available from 2-amino acids,[2] via 2-chloroalkan-1-ols to alkyloxiranes with inversion of configuration at the stereocenter. Thus (R)-methyloxirane is prepared from (S)-alanine, (R)-isopropyloxirane from (S)-valine, (R)-isobutyloxirane from

(S)-leucine, and (S)-sec-butyl-(R)-oxirane from (2S,3S)-isoleucine, respectively. This useful three-step route complements the synthesis of (S)-alkyloxiranes from (S)-2-amino acids via (S)-2-hydroxy acids,[4,5] with retention of configuration at the stereocenter.

The stereoselective conversion of chlorohydrins into diols via oxiranes as intermediates in aqueous potassium hydroxide solution was originally described by Fickett, Garner, and Lucas.[6] In the present procedure, the oxiranes are distilled off as they are formed to prevent subsequent ring-opening. Among different reaction conditions investigated,[7] the procedure given here appears to be most convenient, and is accompanied by almost no racemization (Table II). The enantiomeric purities of the oxiranes are determined directly with high precision by complexation gas chromatography on optically active metal chelates (e.g., Ni(II) bis(2-heptafluorobutyryl-(S)-4-methylthujan-3-onate[8] or Mn(II) bis(3-heptafluorobutyryl-(R)-camphorate,[7] respectively). Depending on the chemical structure of the chloro acids used,[2] the degree of inversion of configuration is less than 0.5% for R = methyl, isopropyl and (S)-sec-butyl, and approximately 1.5% for R = isobutyl. In the latter case, prolonged exposure of the oxirane to the reaction mixture leads to increased racemization. (R)- and (S)-Methyloxirane have been synthesized with retention of configuration from (R)- and (S)-propane-1,2-diol, respectively, by cyclization of the bromoacetates,[9-11] which seems to be superior to the route via bromohydrins.[12] Starting from commercially available (S)-ethyl lactate,[9] other groups have employed different routes to (S)-methyloxirane[13,14] ($[\alpha]_D$ -12.5° (neat)[13] and to (R)-methyloxirane[15,16] ($[\alpha]_D^{24}$ +13.9° (neat),[15] $[\alpha]_D^{20}$ +13.4° (neat),[17] $[\alpha]_D^{22}$ +13.0° (neat),[16] $[\alpha]_D^{25}$ +11.97° (neat)[18]). The apparent deviations of these specific rotations from the maximum optical rotation extrapolated for the pure enantiomer may be ascribed to lack of enantiomeric purity of the substances described, and to inappropriate optical rotation measurements (error in the density, chemical impurities). Enantiomeric impurities in the oxirane can also originate from the starting material since variable fractions of (R)-ethyl lactate (up to 5%) have been detected in commercial (S)-ethyl lactate by gas chromatography on D-Chirasil-Val.[19] The enantiomeric purity of the chiral starting material must be established with certainty in any "chiral pool" transformation.

The "chiral pool" approach appears at present to be superior to other methods of access to optically active alkyl-substituted oxiranes, e.g., enzymatic[20-22] and nonenzymatic[23] epoxidation of prochiral olefins, chromatographic resolution experiments[11,24] and kinetic resolution methods.[8,25,26a] Halohydrins and oxiranes of high enantiomeric purity have recently been obtained by diastereoselective synthesis.[26b] As reviewed previously,[10] a variety of optically active compounds have been synthesized from (R)- and (S)-methyloxirane. Additional examples are macrolides,[27-29] alcohols,[30-32] amino alcohols,[20,33] 1-chloro-2-alkanols[34] and thiiranes.[17,35] The potential of higher, alkyl-substituted oxiranes as building blocks in chiral synthesis awaits its full exploitation. (S)-Ipsenol has been synthesized from (S)-isobutyloxirane,[36,37] which is also available from D-mannitol.[37,38] By stereoselective ring-opening reactions, optically active oligomers (crown ethers)[39] and polymers are conveniently prepared.

TABLE I

(2S)-2-CHLOROALKAN-1-OLS (CHLOROHYDRINS)

$$\begin{array}{c} CH_2OH \\ | \\ Cl-C-H \\ | \\ R \end{array}$$

Substituent R	Yield (%)	bp (°C/mm)	d_4^{20} (g/cm^3)	$[\alpha]_D^{20}$ (°)[a]
$-CH_3$	56	131/725	1.110_1	+17.8
$-CH(CH_3)_2$	70	91/50	1.044_2	+ 3.6
$-CH_2CH(CH_3)_2$	64	92/30	1.005_3	-48.8
(S)-$CH(CH_3)CH_2CH_3$	56	75/10	1.028_6	- 7.6

[a]Rotations were measured on the neat liquids.

TABLE II

(2R)-ALKYLOXIRANES (1,2-EPOXYALKANES)

```
H2C
  | >O
H-C
  |
  R
```

Substituent R	Yield(%)[a]	bp(°C/mm)[b]	e.e.(%)	d_4^{20}(g/cm^3)	$[\alpha]_D^{20}$(°)[c]
$-CH_3$	81/67	34/728	94.6±0.4[d]	0.8309	+13.12
$-CH(CH_3)_2$	93/87	82/730	97.4±0.2[e]	0.8201	- 4.46
$-CH_2CH(CH_3)_2$	84/78	108/730	93.0±0.4[e]	0.8241	+20.47
(S)-$CH(CH_3)CH_2CH_3$	79/73	109/726	97.4±0.2[d,f]	0.7598	+14.4

[a]First number: crude reaction product (organic layer); second number: final yield after redistillation.

[b]Determined in a separate experiment.

[c]Rotations were determined on neat samples; specific rotations are for material of the indicated e.e.

[d]Determined by complexation gas chromatography on Ni(II) bis(2-heptafluorobutyryl-(S)-4-methylthujan-3-onate) (ref 8).

[e]Determined as Mn(II) bis(3-heptafluorobutyryl-(R)-camphorate) (ref 7).

[f]Diastereomeric excess, referring to (2S,3S)-1,2-epoxy-3-methylpentane, (S)-sec-butyl-(R)-oxirane as impurity. Composition: 98.5 ± 0.1% 2R,3S; 1.3 ± 0.1% 2S,3S; 0.2 ± 0.1% 2R,3R, approximately 0% 2S,3R.

1. Institut für Organische Chemie der Universität, Auf der Morgenstelle 18, D-7400 Tübingen, FRG. We thank Mrs. D. Wistuba, University of Tübingen, for determination of the enantiomeric purity of the oxiranes, and Deutsche Forschungsgemeinschaft and Fonds der chemischen Industrie for financial support.
2. Koppenhoefer, B.; Schurig, V. *Org. Synth.* **1987**, *66*, 151.
3. Pachaly, H. in "Ullmanns Encyklopädie der technischen Chemie", Urban & Schwarzenberg: München, Berlin, 1963; Vol. 14, pp 395-397.
4. Tsuji, K.; Hirano, T.; Tsuruta, T. *Makromol. Chem.* **1975**, *Suppl. 1*, 55-70.
5. Mori, K.; Sasaki, M.; Tamada, S.; Suguro, T.; Masuda, S. *Tetrahedron* **1979**, *35*, 1601-1605.
6. Fickett, W.; Garner, H. K.; Lucas, H. J. *J. Am. Chem. Soc.* **1951**, *73*, 5063-5067.
7. Koppenhoefer, B.; Weber, R.; Schurig, V. *Synthesis* **1982**, 316-318.
8. Hintzer, K. Thesis, University of Tübingen, 1983.
9. Golding, B. T.; Hall, D. R.; Sakrikar, S. *J. Chem. Soc., Perkin Trans. 1* **1973**, 1214-1223.
10. Ellis, M. K.; Golding, B. T. *Org. Synth.* **1985**, *63*, 140.
11. Schurig, V.; Koppenhoefer, B.; Buerkle, W. *Angew. Chem., Intern. Ed. Engl.* **1978**, *17*, 937.
12. Price, C. C.; Osgan, M. *J. Am. Chem. Soc.* **1956**, *78*, 4787-4792.
13. Seuring, B.; Seebach, D. *Helv. Chim. Acta* **1977**, *60*, 1175-1181.
14. Schmidt, U.; Gombos, J.; Haslinger, E.; Zak, H. *Chem. Ber.* **1976**, *109*, 2628-2644.
15. Johnston, B. D.; Slessor, K. N. *Can. J. Chem.* **1979**, *57*, 233-235.
16. Hillis, L. R.; Ronald, R. C. *J. Org. Chem.* **1981**, *46*, 3348-3349.

17. Gottarelli, G.; Samori, B.; Moretti, I.; Torre, G. *J. Chem. Soc., Perkin Trans. 2* **1977**, 1105-1111.

18. Utimoto, K.; Uchida, K.; Yamaya, M.; Nozaki, H. *Tetrahedron Lett.* **1977**, 3641-3642.

19. Bayer, E.; Allmendinger, H.; Enderle, G.; Koppenhoefer, B. *Fresenius' Z. Anal. Chem.* **1985**, *321*, 321-324.

20. Bio Research Center Co., Ltd. *Jpn. Kokai Tokkyo Koho JP* 58 65 252 [83 65 252]; *Chem. Abstr.* **1983**, *99*, 193210u.

21. Habets-Crützen, A. Q. H.; Carlier, S. J. N.; de Bont, J. A. M.; Hartmans, S.; Tramper, J.; Wistuba, D.; Schurig, V. *Appl. Microbiol. Biotechn.* in press.

22. Schurig, V.; Wistuba, D. *Angew. Chem., Intern. Ed. Engl.* **1984**, *23*, 796.

23. Kagan, H. B.; Mimoun, H.; Mark, C.; Schurig, V. *Angew. Chem., Intern. Ed. Engl.* **1979**, *18*, 485.

24. Pirkle, W. H.; Rinaldi, P. L. *J. Org. Chem.* **1978**, *43*, 3803-3807.

25. Takeichi, T.; Arihara, M.; Ishimori, M.; Tsuruta, T. *Tetrahedron* **1980**, *36*, 3391-3398.

26. (a) Goguelin, M.; Sepulchre, M. *Makromol. Chem.* **1979**, *180*, 1215-1230; 1231-1241; (b) Oppolzer, W.; Dudfield, P. *Tetrahedron Lett.* **1985**, *26*, 5037-5040.

27. Millar, J. G.; Oehlschlager, A. C.; Wong, J. W. *J. Org. Chem.* **1983**, *48*, 4404-4407.

28. Seidel, W.; Seebach, D. *Tetrahedron Lett.* **1982**, *23*, 159-162.

29. Fink, M.; Gaier, H.; Gerlach, H. *Helv. Chim. Acta* **1982**, *65*, 2563-2569.

30. Voss, G.; Gerlach, H. *Liebigs Ann. Chem.* **1982**, 1466-1477.

31. Kandil, A. A.; Slessor, K. N. *Can. J. Chem.* **1983**, *61*, 1166-1168.

32. Masuda, S.; Nakajima, T.; Suga, S. *Bull. Chem. Soc. Jpn.* **1983**, *56*, 1089-1094.

33. Morrison, J. D.; Grandbois, E. R.; Howard, S. I.; Weisman, G. R. *Tetrahedron Lett.* **1981**, *22*, 2619-2622.

34. Nàjera, C.; Yus, M.; Seebach, D. *Helv. Chim. Acta* **1984**, *67*, 289-300.

35. Schurig, V.; Buerkle, W. *J. Am. Chem. Soc.* **1982**, *104*, 7573-7580.

36. Mori, K. *Tetrahedron* **1976**, *32*, 1101-1106.

37. Rossi, R.; Carpita, A.; Bonaccorsi, F. *Chem. Ind. (Milan)* **1983**, *65*, 694-695.

38. Schmidt, U.; Talbiersky, J.; Bartkowiak, F.; Wild, J. *Angew. Chem., Intern. Ed. Engl.* **1980**, *19*, 198-199.

39. Sato, A.; Hirano, T.; Suga, M.; Tsuruta, T. *Polym. J.* **1977**, *9*, 209-218.

Appendix

Chemical Abstracts Nomenclature (Collective Index Number); (Registry Number)

(R)-Methyloxirane: Propylene oxide, (R)-(+)-(8); Oxirane, methyl-, (R)-(9); (15448-47-2)

(S)-2-Chloropropanoic acid: Propionic acid, 2-chloro-, (S)- (8); Propanoic acid, 2-chloro-, (S)- (9); (29617-66-1)

(S)-2-Chloropropan-1-ol: 1-Propanol, 2-chloro-, (S)-(+)- (8); 1-Propanol, 2-chloro-, (S)- (9); (19210-21-0)

UTILIZATION OF β-CHLORO ALKYLIDENE/ARYLIDENE MALONATES IN ORGANIC SYNTHESIS: SYNTHESIS OF ETHYL CYCLOPROPYLPROPIOLATE

A. $c\text{-}C_3H_5C(=O)CH(CO_2Et)_2$ $\xrightarrow{POCl_3}$ $c\text{-}C_3H_5C(Cl)=C(CO_2Et)_2$

B. $c\text{-}C_3H_5C(Cl)=C(CO_2Et)_2$ $\xrightarrow[2.\ HCl]{1.\ KOH,\ H_2O}$ $c\text{-}C_3H_5C(Cl)=C(CO_2H)CO_2Et$

C. $c\text{-}C_3H_5C(Cl)=C(CO_2H)CO_2Et$ $\xrightarrow[90^{\circ}C]{Et_3N}$ $c\text{-}C_3H_5C{\equiv}CCO_2Et$

Submitted by Osmo Hormi.[1]

Checked by David Oare and Clayton Heathcock.

1. Procedure

A. Diethyl 2-chloro-2-cyclopropylethene-1,1-dicarboxylate. A 1-L, two-necked flask, equipped with magnetic stirrer, reflux condenser and a dropping funnel is charged with 166 g (0.73 mol) of diethyl cyclopropylcarbonylmalonate (Note 1) and 0.5 kg of phosphorus oxychloride (Note 2). The flask is cooled with a water bath, stirring is started and 135 g (0.73 mol) of tributylamine (Note 3) is added from the dropping funnel. The reaction is exothermic. When the addition is complete, the dropping funnel is replaced by a glass stopper and the water bath is replaced by an oil bath. The mixture is heated at 110°C with stirring for 5-6 hr.

Excess phosphorus oxychloride is removed as well as possible with a rotary evaporator under reduced pressure. The residue is cooled to room temperature and 300 mL of diethyl ether is added. The mixture is poured into a separatory funnel. Hexane is added until the two phases separate cleanly and the funnel is shaken vigorously. The phases are separated and the lower layer is extracted with three 250-mL portions of ether (Note 4). The combined organic layers are washed with 300 mL of cold aqueous 10% hydrochloric acid and 200 mL of aqueous 5% sodium hydroxide (Note 5) and then concentrated carefully with a rotary evaporator to give 136-156 g (70-87%) of crude diethyl 2-chloro-2-cyclopropylethene-1,1-dicarboxylate.[2]

B. The crude chloromalonate is dissolved in 100 mL of 95% ethanol and transferred to a 1-L, round-bottomed flask equipped with a magnetic stirring bar. Stirring is begun and a solution of potassium hydroxide in 350 mL of 95% ethanol (0.0035 mol of potassium hydroxide per gram of chloromalonate) (Note 6) is added dropwise from an addition funnel. A slightly exothermic reaction is noted. After the addition is complete, the mixture is stirred for 3 hr (or until the mixture is neutral to litmus, Note 7). Excess ethanol is removed with a rotary evaporator under reduced pressure and the residue is dissolved in 300 mL of water and extracted with 350 mL of ether (Note 8). The phases are separated and some ice is added to the aqueous phase. The cooled, aqueous phase is acidified with concentrated hydrochloric acid and extracted with three 300-mL portions of ether. The ether phase is dried with anhydrous sodium sulfate, filtered, and concentrated under reduced pressure with a rotary evaporator to give 72-94 g (70-80%) of crude monoester.

C. *Ethyl cyclopropylpropiolate*. The crude product is transferred to a 500-mL, round-bottomed flask equipped with a magnetic stirring bar and a reflux condenser. A solution of 0.70 mL of triethylamine (Note 9) per gram of crude monoester from part B in about 200 mL of toluene is added. The mixture is heated using an oil bath at 90°C with stirring until the evolution of carbon dioxide has subsided, and is then heated for another hour (Note 10). The mixture is cooled to room temperature, washed with 300 mL of aqueous 10% hydrochloric acid (Note 11) and finally with 300 mL of aqueous 5% sodium carbonate. The organic layer is dried with anhydrous sodium sulfate, filtered, and concentrated under reduced pressure with a rotary evaporator. Fractional distillation of the residue gives the product, 87-95°C at 10 mm. The yield of the final step is 30-46 g (66-78%); the overall yield is 33-54% (Note 12).

2. Notes

1. Diethyl cyclopropylcarbonylmalonate is prepared by the procedure of Price and Tarbell[3a] or Reynolds and Hauser.[3b] On scale-up the checkers found that a slight modification is necessary: The procedure of Price and Tarbell was used:[3a] 45 g (1.9 mol) of magnesium turnings, 1 mL of carbon tetrachloride, and approximately 20 mL of a solution of 281 mL (269 g, 1.9 mol) of diethyl malonate in 148 mL of absolute ethanol were combined. After the reaction begins, the addition of diethyl malonate solution is completed so that the reaction is maintained at a fairly vigorous rate. If the reaction subsides prior to the completion of the addition of the diethyl malonate solution, the addition is interrupted and a portion (300 to 400 mL) of the specified 550 mL of dry ether is added cautiously until the reaction resumes, whereupon the addition of the diethyl malonate solution is resumed. After the remainder of the diethyl malonate solution has been added and the reaction mixture cooled, the remaining dry ether is added cautiously and the mixture is worked-up as described[3a] using 350 mL of dry benzene. The residue is

dissolved in 550 mL of dry ether and treated as described by Reynolds and Hauser[3b] with 168 mL (194 g, 1.9 mol) of cyclopropanecarboxylic acid chloride in 230 mL of dry ether. After the addition is completed, an additional 40 mL of dry ether is added and the mixture is cooled and worked up as described[3b] using 2L of aqueous 25% sulfuric acid, 800 mL of ether, 600 mL of saturated aqueous sodium bicarbonate, 200 mL of water, and 100 mL of brine. The crude product is distilled (85-95°C; 0.05 mm) and 362 g (1.6 mol, 86%) of a clear liquid is obtained. ^{1}H NMR δ: 1.01 (m, 2 H), 1.20 (m, 2 H), 1.31 (t, 6 H, J = 7.1), 2.12 (tt, 1 H, J = 4.5, 7.8), 4.28 (q, 4 H, J = 7.1), 4.58 (s, 1 H).

2. Commercial phosphorus oxychloride was used without purification.

3. Commercial tributylamine was used without purification.

4. Extraction with a mixture of ether and hexane is repeated until ether and the lower layer readily separate.

5. Tributylamine is liberated from the lower layer by addition of sodium hydroxide.

6. Commercial potassium hydroxide (min 85.5% of potassium hydroxide) was used. The yield is based on potassium hydroxide.

7. Dilution of 1 mL of the mixture in 5 mL of water gave pH 7-8.

8. The ether phase is dried with anhydrous sodium sulfate, filtered and concentrated with a rotary evaporator to give 32-46 g of recovered starting material.

9. Commercial triethylamine was used without purification.

10. The checkers found that this process requires approximately 24 hr.

11. Triethylamine is liberated from the water phase by addition of sodium hydroxide.

12. Cyclopropylpropiolic acid ethyl ester has the following spectra: ^{1}H NMR (CCl_4) δ: 0.84 and 0.93 (4 H, ring CH_2), 1.25 (t, 3 H, CH_3 ester), the ring-CH is hidden under the ester CH_3-triplet, 4.05 (q, 2 H, CH_2 ester); IR (CCl_4) cm^{-1}: 2220 (C≡C), m, 1710 (C=O), s, 1255 (C-O-C), s, 1030-1040 two bands (cyclopropyl), m, and 860-880 two bands (cyclopropyl), w; MS: M (calculated for $C_8H_{10}O_2$) 138.068, M^+ is not readily detected, 94 (20%), 93 (100%), 66 (65%), 65 (53%), 63 (10%), 53 (15%), 40 (10%).

3. Discussion

Substituted cyclopropyl rings conjugated with a triple bond system have recently received attention as C_5 building blocks.[4] The procedure described here is a modification of the decarboxylation-elimination reaction for the preparation of α,β acetylenic acids from enol sulfonates of acyl malonates.[5,6,7] Addition of aqueous alkali to the enol sulfonate of diethyl cyclopropylcarbonylmalonate gives cyclopropylpropiolic acid, but the yield is low.

The major advantages of this procedure over the enol sulfonate procedure lie in the availability of diethyl 2-chloro-2-cyclopropylethene-1,1-dicarboxylate from the corresponding acylmalonate and phosphorus oxychloride, and the fast, homogeneous, decarboxylative elimination reaction of the triethylamine salt of the half-ester in dry organic solvents. The conditions described here, with slight modifications (overnight treatment), have been used for a variety of β-chloro alkylidene/arylidene malonates as shown in Table I.

TABLE I

α,β-ACETYLENIC ESTERS FROM β-CHLORO ALKYLIDENE MALONATES

$$\begin{matrix} Cl \\ \quad \searrow \\ \quad\quad C{=}C(COOH)COOEt \\ \quad \nearrow \\ R \end{matrix} + N(Et)_3 \longrightarrow R{-}C{\equiv}C{-}COOEt + HCl\cdot N(Et)_3 + CO_2$$

R	Yield
Phenyl	80%
2-Thienyl	90%
p-NO_2-C_6H_4	70%
Isopropyl	67%
tert-Butyl	70%

Sometimes the acetylenic ester rearranges to the corresponding allenic ester. For example, when the triethylamine salt of 3-chloro-2-ethoxycarbonyl-4-phenyl-2-hexenoic acid is refluxed in toluene, the allenic ester and acetylenic ester are obtained in a ratio of 3:7 (total yield 70%). There are alternative routes to cyclopropylpropiolic acids and esters, such as adding butyllithium to corresponding acetylenes and treating the product with carbon dioxide or methyl chloroformate.[4]

1. Institutionen for Organisk Kemi, Åbo Akademi, Akademigatan 1, SF-20500 Åbo 50, Finland.

2. Cf. Friedrich, K.; Thieme, H. K. *Chem. Ber.* **1970**, *103*, 1982; Friedrich, K.; Thieme, H. K. *Synthesis* **1973**, 111.

3. (a) Price, J. A.; Tarbell, D. S. *Org. Synth., Collect. Vol. 4* **1963**, 285; (b) Reynolds, G. A.; Hauser, C. R. *Org. Synth., Collect. Vol. 4* **1963**, 708.

4. Bengtson, G.; Keyaniyan, S.; de Meijere, A. *Chem. Ber.* **1986**, 119, 3607, and references therein.

5. Fleming, I.; Harley-Mason, J. *J. Chem. Soc.* **1963**, 4771 and 4778.

6. Brown, E. J. D.; Harley-Mason, J. *J. Chem. Soc. (C)* **1966**, 1390.

7. Fleming, I.; Owen, C. R. *J. Chem. Soc. (C)* **1971**, 2013.

Appendix

Chemical Abstracts Nomenclature (Collective Index Number); (Registry Number)

Diethyl cyclopropylcarbonylmalonate: Malonic acid, (cyclopropylcarbonyl)-, diethyl ester (8); Propanedioic acid, (cyclopropylcarbonyl)-, diethyl ester (9): (7394-16-3)

Diethyl malonate: Malonic acid, diethyl ester (8); Propanedioic acid, diethyl ester (9); (105-53-3)

OXIDATIVE CLEAVAGE OF AN AROMATIC RING: cis,cis-MONOMETHYL MUCONATE FROM 1,2-DIHYDROXYBENZENE

(2,4-Hexadienedioic acid, monomethyl ester, (Z,Z)-)

OH OH $\xrightarrow[\text{pyridine}]{O_2/CuCl/CH_3OH}$ CO_2H CO_2CH_3

Submitted by Donald Bankston.[1]

Checked by Won Hun Ham and Leo A. Paquette.

1. Procedure

A. cis,cis-Monomethyl muconate. A 1000-mL, three-necked, round-bottomed flask is equipped with a mechanical stirrer and an addition funnel (Note 1, Figure 1). The flask is charged with 400 mL of pyridine, 5 mL (0.12 mol) of methanol, and 9.9 g (0.10 mol) of cuprous chloride under an atmosphere of nitrogen (Note 2). The resultant yellow solution is stirred vigorously at room temperature until the cuprous chloride dissolves (Note 3). The nitrogen is removed and oxygen is bubbled into the flask below the surface of the liquid for approximately 30 min (Note 4). A solution, composed of 1000 mL of pyridine, 5.5 g (0.05 mol) of 1,2-dihydroxybenzene (catechol), and 5 mL (0.12 mol) of methanol, is degassed and slowly added to the flask over a 2-hr period with efficient stirring (Note 5). The reaction mixture is stirred for an additional 30 min before the pyridine is removed at reduced pressure. The dark green residue is dissolved in 300 mL of ethyl ether (Note 6) and

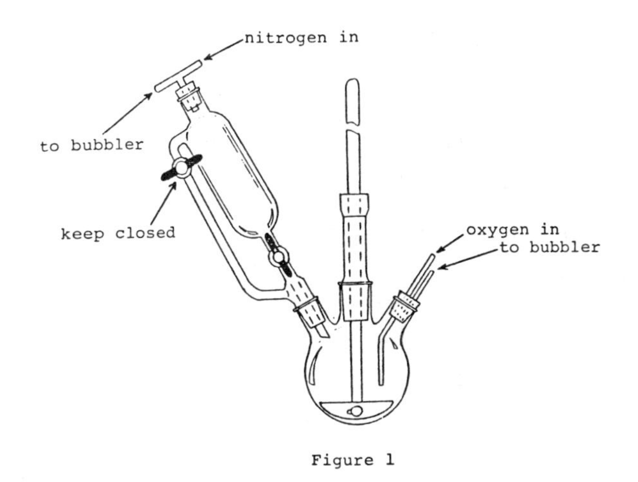

Figure 1

300 mL of 6 N hydrochloric acid; this solution is stirred for 10 min, filtered over Celite, and the organic layer is isolated. The ether is removed at reduced pressure and the resultant brown residue is boiled sequentially with six 50-mL portions of hexane (Note 7). The hot liquid is carefully decanted, leaving behind most of the dark residue. The yellow to colorless solid is crystallized from hot hexane (or methanol) to yield 5.6-6.2 g (71-80%) of colorless needles, mp 80-81°C (Note 8).

2. Notes

1. The checkers employed an alternate device of the following type and

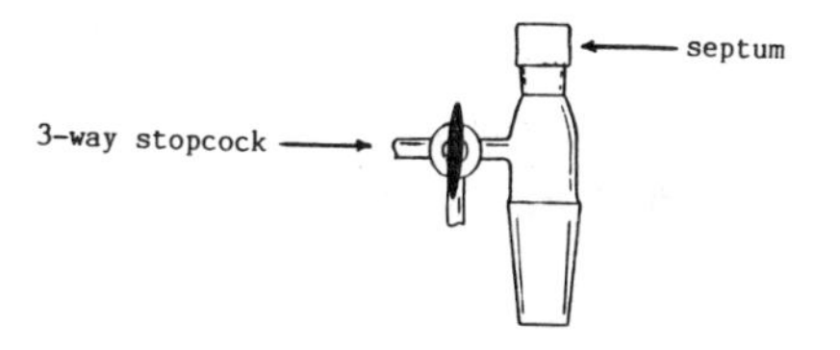

introduced the methanolic catechol solution subsequently via syringe.

2. Pyridine (Mallinckrodt Inc., reagent grade) was distilled over potassium hydroxide, bp 114-116°C. Cuprous chloride was prepared fresh,[2] washed with anhydrous methanol (distilled from magnesium metal), and dried under reduced pressure. Methanol (Fisher Scientific Company, purified grade) was dried over molecular sieves. The concentration of the methanol should be within the range of 3-8 mol/mol of catechol[3,4] and anhydrous conditions are necessary.[4]

3. Dissolution requires 1.5-2 hr. If the solution is not properly degassed, it will turn green prematurely. The green color indicates that oxygen absorption by the cuprous chloride-pyridine complex has occurred, but it also means that any undissolved cuprous chloride has been oxidized. Therefore, nitrogen should be bubbled into the flask at a brisk rate and stirring should not commence until addition of the cuprous chloride is complete.

4. As oxygen is introduced into the flask the solution becomes dark green and slightly viscous.

5. Catechol was purchased from Fisher Scientific Company (resublimed). The addition funnel should be charged with a nitrogen atmosphere throughout addition to obviate oxidation of the catechol.

6. Ethyl ether was purchased from Mallinckrodt Inc. (analytical grade); methylene chloride may be substituted, but the monomethyl ester is more soluble in ether. At least two other extractions will be necessary to optimize the yield when methylene chloride is used.

7. Hexane was purchased from Fisher Scientific Company (technical grade).

8. The spectral properties are as follows: ^{1}H NMR (300 MHz, CCl_4) δ: 3.74 (s, 3 H); 5.88-6.09 (m, 2 H); 7.75-8.26 (m, 2 H); 11.29 (s, 1 H).

3. Discussion

Phenol also undergoes oxidative cleavage in the presence of O_2/CuCl/pyridine and methanol to give cis,cis-monomethyl muconate, but the yield is not high.[4,5] It has also been observed that catechol, under anaerobic conditions, reacts with cupric methoxy chloride in pyridine containing water and methanol to give cis,cis-monomethyl muconate in high yield.[6] Catechol and phenol may also be converted to cis,cis-muconic acid by a metal-catalyzed peracetic acid oxidation;[7] subsequently, treatment with diazomethane gives the monomethyl or dimethyl ester.

1. Laboratory of J. C. Barborak, Department of Chemistry, University of North Carolina at Greensboro, Greensboro, NC 27412.
2. Keller, R. N.; Wycoff, H. D. *Inorg. Synth.* **1946**, *2*, 1-3.
3. Tsuji, J.; Takayanagi, H. *J. Am. Chem. Soc.* **1974**, *96*, 7349-7350.
4. Tsuji, J.; Takayanagi, H. *Tetrahedron* **1978**, *34*, 641-644.
5. Tsuji, J.; Takayanagi, H. *Tetrahedron Lett.* **1976**, 1365-1366.
6. Rogic, M. M.; Demmin, T. R.; Hammond, W. B. *J. Am. Chem. Soc.* **1976**, *98*, 7441-7443.
7. Pandell, A. J. *J. Org. Chem.* **1976**, *41*, 3992-3996.

Appendix

Chemical Abstracts Nomenclature (Collective Index Number); (Registry Number)

cis,cis-Monomethyl muconate: 2,4-Hexadienedioic acid, monomethyl ester, (Z,Z)- (9); (61186-96-7)

1,2-Dihydroxybenzene: Pyrocatechol (8); 1,2-Benzenediol (9); (120-80-9)

Methanol (8,9); (67-56-1)

Cuprous chloride: Copper chloride (8); Copper chloride (CuCl) (9); (7758-89-6)

PREPARATION OF 2-PROPYL-1-AZACYCLOHEPTANE

FROM CYCLOHEXANONE OXIME

(1H-Azepine, hexahydro-2-propyl-, (±))

A. Cyclohexanone oxime → (CH_3SO_2Cl, NEt_3, CH_2Cl_2) → cyclohexanone oxime O-SO_2CH_3

B. Cyclohexanone oxime O-SO_2CH_3 → (Pr_3Al, CH_2Cl_2, Hexane) → [2-propyl-azacycloheptene] → (i-Bu_2AlH, Hexane) → 2-propyl-1-azacycloheptane

Submitted by Keiji Maruoka, Shuichi Nakai, and Hisashi Yamamoto.[1]

Checked by Jeffrey Doney and Clayton H. Heathcock.

1. Procedure

Caution! Part B of this procedure should be carried out in a well-ventilated hood to prevent exposure to methanethiol, a side-product.

A. Cyclohexanone oxime methanesulfonate. A dry, 1-L, two-necked, round-bottomed flask is equipped with a gas inlet, rubber septum, and magnetic stirring bar. The flask is charged with 17.0 g (0.15 mol) of cyclohexanone oxime (Note 1) and flushed with argon, after which 300 mL of dichloromethane followed by 25 mL (0.18 mol) of triethylamine (Note 2) are injected through the septum into the flask. The solution is stirred and cooled to a temperature of -15 to -20°C in a dry ice-carbon tetrachloride bath, while 12.8 mL

(0.165 mol) of methanesulfonyl chloride is added over a 20-min period (Notes 3 and 4). The resulting mixture is stirred at this temperature for 15 min, and poured into 300 mL of ice-water in a 1-L separatory funnel with the aid of three 30-mL portions of dichloromethane to rinse the flask. The lower organic layer is separated, and the aqueous layer is extracted with a 50-mL portion of dichloromethane. The combined extracts are washed successively with 250 mL of cold aqueous 10% hydrochloric acid, 250 mL of saturated sodium bicarbonate, 250 mL of brine, dried over anhydrous sodium sulfate, and concentrated with a rotary evaporator at room temperature to give 27.2-28.8 g of crude solid cyclohexanone oxime methanesulfonate (Note 5). This material is used in part B without purification (Note 6).

B. 2-Propyl-1-azacycloheptane. A dry, 2-L, three-necked, round-bottomed flask is equipped with a variable speed mechanical stirrer, 300-mL pressure-equalizing dropping funnel bearing a gas inlet at its top, and a rubber septum. The apparatus is flushed with argon, after which 243 mL of hexane (Note 7) and 57 mL (0.3 mol) of tripropylaluminum (Note 8) are injected through the septum into the flask. The solution is stirred and cooled to a temperature of -73 to -78°C in a dry ice-methanol bath. The crude cyclohexanone oxime methanesulfonate prepared in Part A is dissolved in 100 mL of dichloromethane, transferred to the dropping funnel, and added to a 1 M solution of tripropylaluminum in hexane over a 30-min period (Note 9). The mixture is allowed to warm to 0°C, stirred for 1 hr, and 225 mL (0.225 mol) of a 1 M solution of diisobutylaluminum hydride in hexane (Note 10) is added at 0°C and the mixture is further stirred at 0°C for 1 hr (Note 11). After addition of 100 mL of dichloromethane and 88.2 g (2.1 mol) of sodium fluoride, 28.4 mL (1.58 mol) of water is injected dropwise at 0°C (Note 12). Vigorous stirring of the resulting suspension is continued for 30 min at room temperature, and the contents of the flask are filtered with five 30-mL portions of dichloromethane (Note 13). The combined filtrates are evaporated under reduced pressure with a rotary evaporator. Distillation of the residual liquid under reduced pressure affords 11.3-12.2 g (53-58%) of 2-propyl-1-azacycloheptane as a colorless liquid, bp 79-81°C (18 mm) (Notes 14 and 15).

2. Notes

1. Reagent-grade cyclohexanone oxime, purchased from Wako Pure Chemical Industries, Ltd. (Japan), was used as received. The checkers used material obtained from the Aldrich Chemical Company, Inc. A suitable material may be prepared according to the procedures in *Organic Syntheses*.[2]

2. Both reagent-grade dichloromethane and triethylamine were dried and stored over Linde type 4 Å molecular sieves.

3. The solution turns to a white suspension after half of the methanesulfonyl chloride is added; methanesulfonyl chloride, available from Tokyo Kasei Kogyo Company, Ltd. (Japan), was used without any purification.

4. The checkers found that an addition time of 40 min is required to maintain the temperature of the reaction mixture below -15°C.

5. If a crude oil was obtained at this stage, it can be solidified by cooling.

6. The reaction in Part A proceeds in almost quantitative yield.[3] Accordingly the crude cyclohexanone oxime methanesulfonate can be used without any purification. Prolonged standing at room temperature may cause serious decomposition. The crude material may be stored in a freezer, or as a dichloromethane solution in a refrigerator, and can be recrystallized from ether-hexane to give the white solid (mp 43-45°C).[4]

7. Reagent-grade hexane was dried and stored over sodium.

8. Neat tripropylaluminum of 96% purity was supplied in a metal cylinder from Toyo Stauffer Chemical Company, Ltd. (Japan). This reagent is contaminated by 1.2% of triethylaluminum, 2.2% of triisobutylaluminum, and other compounds. Neat tripropylaluminum is also available from Aldrich Chemical Company, Inc. Since neat tripropylaluminum is pyrophoric and reacts violently with oxygen and water, the syringe should be washed with hexane immediately after addition.

9. The checkers found that an addition time of 60 min is required to maintain the temperature of the reaction mixture below -73°C.

10. Diisobutylaluminum hydride in hexane was available from Aldrich Chemical Company, Inc. and Kanto Chemical Company, Inc. (Japan).

11. Methanethiol is generated as a side-product by the reduction of the methanesulfonate with diisobutylaluminum hydride.

12. To avoid excessive foaming at the beginning of the hydrolysis water should be added carefully by syringe. The rate of addition may be increased once the initially vigorous foaming subsides.

13. The sodium fluoride-water work-up offers an excellent method for large-scale preparations, and is generally applicable for product isolation in the reaction of organoaluminum compounds.[5]

14. The elemental analysis and the spectral properties of the product are as follows: Anal. Calcd for $C_9H_{19}N$: C, 76.61; H, 13.47; N, 9.92. Found: C, 76.75; H, 13.74; N, 9.51; IR (liquid film) cm^{-1}: 3320, 2860-2970, 1460, 1165; ^1H NMR ($CDCl_3$) δ: 0.87-0.94 (3 H, m, CH_3), 1.23-1.82 (13 H, m), 2.54-2.73 (2 H, m, CH_2N), 2.96-3.03 (1 H, m, CHN). A boiling point of 193-194°C at 750 mm has been reported for 2-propyl-1-azacycloheptane.[6]

15. Gas chromatographic analysis using a 25-m PEG-HT capillary column at 80°C indicated a purity of 97% (retention time: 8.1 min) based on tripropylaluminum of 96% purity (Note 8). The unrearranged product, cyclohexylpropylamine (retention time: 6.9 min), was less than 1%.

3. Discussion

This procedure illustrates a new, general method for the one-nitrogen ring expansion of cyclic ketoximes leading to α-alkylated, cyclic, secondary amines.[4,7] The key step in the sequence is the organoaluminum-promoted Beckmann rearrangement of ketoxime derivatives, in which the organoaluminum compounds are used as amphiphilic reagents to induce the Beckmann rearrangement of oxime derivatives as well as to capture the intermediary imino carbocation by the alkyl group that is originally attached to aluminum. The conventional process for accomplishing this transformation consists of the following steps: (i) Beckmann rearrangement of ketoxime or its derivative to lactam; (ii) conversion of the lactam to imino ether using trialkyloxonium tetrafluoroborate; (iii) alkylation of the imino ether with alkyllithium or Grignard reagent to produce imine, which requires a considerably longer time for execution.[8]

As oxime derivatives, oxime sulfonates can be used preferentially for the following reasons: (1) they are readily available from oximes using p-toluenesulfonyl chloride or methanesulfonyl chloride in the presence of base in almost quantitative yield; (2) they are easy to handle because of their fine crystalline properties; (3) they are sufficiently reactive to initiate the rearrangement by organoaluminum reagents.

As shown in Table I, this reaction sequence has wide generality and is readily applicable to the straightforward synthesis of various naturally occurring alkaloids such as coniine,[9] pumiliotoxin C,[10] and solenopsin A and B.[11] Oxime sulfonates of either linear or cyclic structures may be used. Obviously, the regioselectivity of the reaction follows the general rule of the Beckmann rearrangement,[12] and preferential migration of the group anti to the oxime sulfonate is observed. Diethylaluminum alkynides can be successfully used for the selective introduction of alkynyl groups to a substrate in preference to an ethyl group. Furthermore, the present procedure reduces the intermediate imine directly without isolation by using diisobutylaluminum hydride, thus excluding the troublesome isolation of unstable cyclic imino compound.

The organoaluminum-promoted Beckmann rearrangement-alkylation sequence represents a modern aspect of the classical Beckmann rearrangement, and has proved effective with other aluminum reagents of type R_2AlX (X = SR, SeR, and CN) which would function in a similar way to trialkylaluminum compounds. Thus, a series of imino thioethers, selenoethers, and nitriles can be prepared with rigorous regioselectivity by using organoaluminum thiolates, selenolates, and cyanide, respectively.[4]

2-Propyl-1-azacycloheptane has been prepared by reduction of 2-aza-1-oxo-3-propylcycloheptane with lithium aluminum hydride,[6] and from azacycloheptane by conversion to its formamidine, alkylation with 1-iodopropane, and subsequent hydrazinolysis.[13]

TABLE I

PREPARATION OF α-ALKYLATED AMINES FROM OXIME SULFONATES WITH TRIALKYLALUMINUM - DIISOBUTYLALUMINUM HYDRIDE

Oxime Sulfonate (mp, °C)	Trialkylaluminum	Amine	Yield (%)
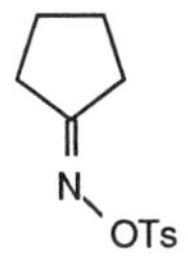 (75-77)	Pr_3Al[a]	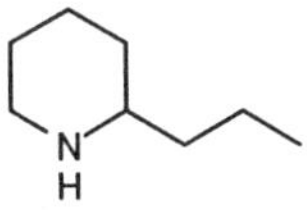	55-58
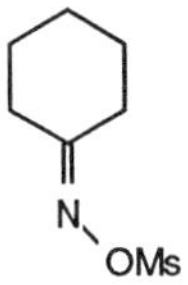 (43-45)	Me_3Al $Et_2AlC{\equiv}CBu$	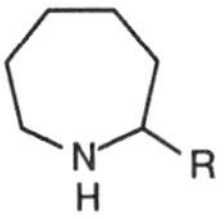	70 (R = Me) 67 (R = C≡CBu)
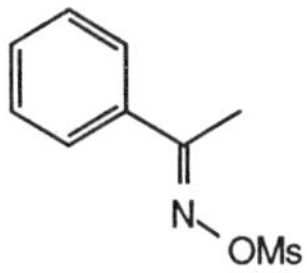 (64-65)	Me_3Al $Et_2AlC{\equiv}CBu$ $Et_2AlC{\equiv}CPh$	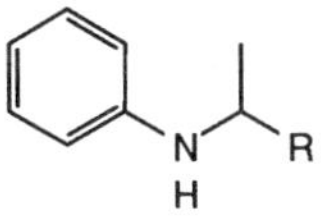	67 (R = Me) 83 (R = C≡CBu) 67 (R = C≡CPh)
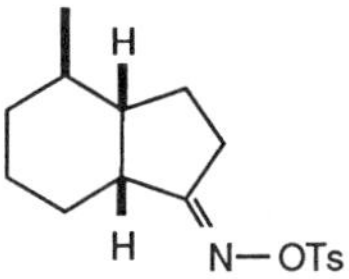 (67-71)	Pr_3Al[b]	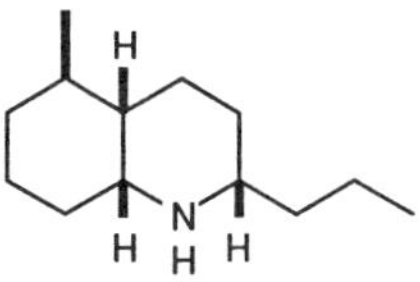 Pumiliotoxin C	60

[a]Treatment with Pr_3Al at 40-80°C for 15-30 min.

[b]Treatment with Pr_3Al at 25°C for 30 min.

1. Department of Applied Chemistry, Faculty of Engineering, Nagoya University, Chikusa, Nagoya 464, Japan.

2. Eck, J. C.; Marvel, C. S. *Org. Synth., Collect. Vol. II* **1943**, 76; Bousquet, E. W. *Org. Synth., Collect. Vol. II* **1943**, 313.

3. Crossland, R. K.; Servis, K. L. *J. Org. Chem.* **1970**, *35*, 3195.

4. Maruoka, K.; Miyazaki, T.; Ando, M.; Matsumura, Y.; Sakane, S.; Hattori, K.; Yamamoto, H. *J. Am. Chem. Soc.* **1983**, *105*, 2831.

5. Yamamoto, H.; Maruoka, K. *J. Am. Chem. Soc.* **1981**, *103*, 4186.

6. Ehrhart, G.; Seidl, G. *Chem. Ber.* **1964**, *97*, 1994.

7. Hattori, K.; Matsumura, Y.; Miyazaki, T.; Maruoka, K.; Yamamoto, H. *J. Am. Chem. Soc.* **1981**, *103*, 7368.

8. Oppolzer, W.; Fehr, C.; Warneke, J. *Helv. Chim. Acta* **1977**, *60*, 48.

9. Chemnitius, F. *J. Prakt. Chem.* **1928**, *118*, 25.

10. Inubushi, Y.; Ibuka, T. *Heterocycles* **1977**, *8*, 633.

11. Matsumura, Y.; Maruoka, K.; Yamamoto, H. *Tetrahedron Lett.* **1982**, *23*, 1929.

12. Donaruma, I. G.; Heldt, W. Z. *Org. React.* **1960**, *11*, 1.

13. Meyers, A. I.; Edwards, P. D.; Rieker, W. F.; Bailey, T. R. *J. Am. Chem. Soc.* **1984**, *106*, 3270.

Appendix

Chemical Abstracts Nomenclature (Collective Index Number); (Registry Number)

2-Propyl-1-azacycloheptane: 1H-Azepine, hexahydro-2-propyl-, (±)- (11); (85028-29-1)

Cyclohexanone oxime methanesulfonate: Cyclohexanone, O-(methylsulfonyl)oxime (10); (80053-69-6)

Cyclohexanone oxime (8); Cyclohexanone, oxime (90: (100-64-1)

Methanesulfonyl chloride (8,9); (124-63-0)

Tripropylaluminum: Aluminum, tripropyl- (8,9); (102-67-0)

Diisobutylaluminum hydride: Aluminum, hydrodiisobutyl- (8); Aluminum, hydro-bis(2-methylpropyl)- (9); (1191-15-7)

6-DIETHYLPHOSPHONOMETHYL-2,2-DIMETHYL-1,3-DIOXEN-4-ONE

(Phosphonic acid, [(2,2-dimethyl-4-oxo-4H-1,3-dioxin-6-yl)methyl])-,

diethyl ester)

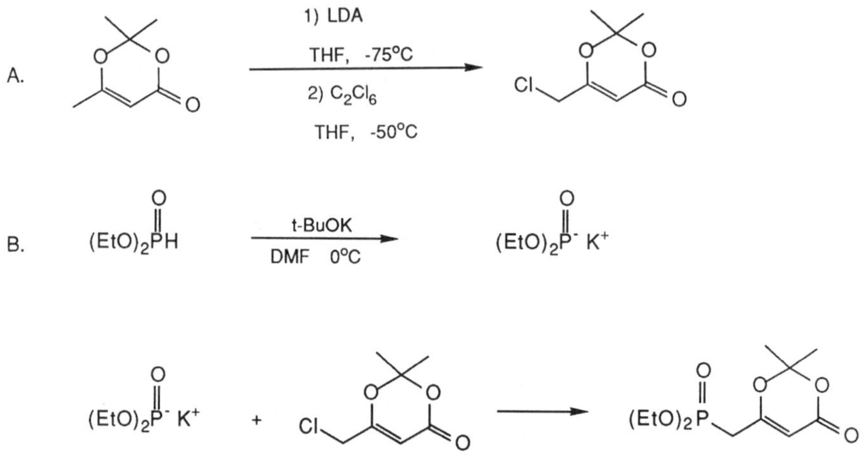

A. 1) LDA, THF, -75°C; 2) C_2Cl_6, THF, -50°C

B. $(EtO)_2P(O)H$ → t-BuOK, DMF 0°C → $(EtO)_2P(O)^- K^+$

$(EtO)_2P(O)^- K^+$ + Cl → $(EtO)_2P(O)$

Submitted by Robert K. Boeckman, Jr., Robert B. Perni, James E. Macdonald, and Anthony J. Thomas.[1]

Checked by Stacey C. Slater and James D. White.

1. Procedure

A. *6-Chloromethyl-2,2-dimethyl-1,3-dioxen-4-one*. A three-necked, 500 mL, round-bottomed flask is fitted with a nitrogen inlet, rubber septum, and a 125-mL dropping funnel. The flask is flame-dried, flushed with nitrogen and charged with diisopropylamine (22.0 mL, 0.16 mol) and 100 mL of tetrahydrofuran (THF) (Note 1). This solution is cooled in an ice bath, and the dropping funnel is charged with a solution of butyllithium (80.0 mL of a 1.88 M solution in hexane, 0.15 mol) which is added dropwise over 15 min (Note

2). The resulting solution is cooled to approximately -75°C in a dry ice-acetone bath and treated with a solution of 2,2,6-trimethyl-1,3-dioxen-4-one[2] (16.0 g, 0.11 mol) in tetrahydrofuran (20 mL) dropwise over 20 min (Note 3). During the addition, a fine yellow suspension forms. The enolate solution is stirred at -75°C for an additional 15 min and then transferred via cannula to a 1-L flask containing hexachloroethane[3] (39.0 g, 0.16 mol) (Note 4) in tetrahydrofuran (150 mL) at -50°C to -55°C (dry ice-acetone bath) over 30 min. When the addition is complete, any residual enolate is transferred with an additional portion of tetrahydrofuran (20 mL). The resulting reaction mixture is allowed to warm slowly to -25°C over 30 min and poured into ice-cold aqueous 10% hydrochloric acid (200 mL), and the mixture is briefly shaken to discharge the red color. The organic layer is separated, and the aqueous layer is extracted with ether (2 x 100 mL). The combined organic extracts are washed with saturated aqueous sodium bicarbonate solution (100 mL), saturated aqueous sodium chloride solution (100 mL), dried over sodium sulfate, and concentrated under reduced pressure to afford 31.50-35.50 g of an oily solid. Column chromatography on Florisil (100-200 mesh, 400 g) (Note 5) and elution with hexane (1 L) and 20% ethyl acetate-hexane (2 L), gives 12.17-12.67 g (63-65%) of the desired product as a yellow oil (Notes 6, 7, and 8).

B. 6-Diethylphosphonomethyl-2,2-dimethyl-1,3-dioxen-4-one. A 500-mL, three-necked flask is outfitted as above, flushed with nitrogen and charged with potassium tert-butoxide (21.0 g, 0.187 mol) and dimethylformamide (200 mL) (Note 9). The stirring mixture is cooled in an ice bath and treated with diethyl phosphite (26.7 g, 0.193 mol). The resulting solution is stirred in the ice bath for 20-40 min and then treated dropwise with a solution of 6-chloromethyl-2,2-dimethyl-1,3-dioxen-4-one (11.00 g, 0.062 mol) in tetrahydrofuran (50 mL) over 20 min. The resulting purple solution is stirred an additional 15 min at 0°C and treated with concentrated hydrochloric acid dropwise until the purple color is discharged (approx. 6 mL). The resulting mixture is filtered by suction through Celite (Note 10), and the collected solids are washed with tetrahydrofuran (50 mL). The combined organic portions are treated with several grams of anhydrous potassium carbonate, filtered, and the tetrahydrofuran is removed with a rotary evaporator. Dimethylformamide and excess diethyl phosphite are removed by distillation at 0.4 mm with the bath temperature maintained below 50°C (Note 11). The residue is diluted with ethyl acetate (200 mL) and placed in the refrigerator at 0°C overnight. The solid which precipitates is removed by filtration, and the filtrate is concentrated under reduced pressure to ~ 75 mL and purified by flash chromatography (Notes 12 and 13) on 700 g of Florisil (9 x 22 cm column). Elution with 3 L of 1:1 ethyl acetate-hexane, 3 L of 3:1 ethyl acetate-hexane, and then 3 L of 100% ethyl acetate affords 8.30-8.56 g (48-50%) of 6-diethylphosphonomethyl-2,2-dimethyl-1,3-dioxen-4-one (Notes 14 and 15). Mixed fractions may be rechromatographed to afford an additional 2-4% of product.

2. Notes

1. Tetrahydrofuran was distilled under a nitrogen atmosphere from sodium benzophenone ketyl. Diisopropylamine was distilled under a nitrogen atmosphere from calcium hydride.

2. A solution of butyllithium in hexane (~ 1.8 M) was obtained from Lithcoa and standardized by titration against 2,5-dimethoxybenzyl alcohol.[4]

3. 2,2,6-Trimethyl-1,3-dioxen-4-one is commerically available from the Aldrich Chemical Company, Inc. and may be used without further purification.

4. Hexachloroethane was obtained from the Aldrich Chemical Company, Inc. and used without further purification.

5. Florisil is a magnesium silicate adsorbent obtained from the Floridin Company.

6. Substantial amounts of unreacted hexachloroethane may be recovered from early fractions.

7. The reaction may be carried out equally well on a 32-g scale.

8. The NMR and IR spectral data of the chloride are as follows: ^{1}H NMR ($CDCl_3$) δ: 1.74 (s, 6 H), 4.02 (s, 2 H), 5.53 (s, 1 H), IR (film) cm^{-1}: 2970, 1730, 1640, 1390, 1280, 1210, 1024.

9. Dimethylformamide (DMF) was distilled under reduced pressure (20 mm) from calcium hydride. Diethyl phosphite may be used directly from a freshly-opened bottle or redistilled before use.

10. This filtration is very slow, and a wide sintered-glass funnel is recommended.

11. The product phosphonate decomposes to diethylphosphonoacetone above 50°C, and care must be taken during the distillation and concentration of chromatography fractions that heating baths do not exceed this temperature.

12. The procedure of W. C. Still was utilized.[5]

13. Chromatographic fractions were analyzed by TLC by elution with ethyl acetate, and were visualized with a permanganate spray. The phosphonate had R_f = 0.35.

14. The NMR and IR spectral data of the phosphonate are as follows: ^{1}H NMR ($CDCl_3$) δ: 1.37 (t, 6 H), 1.72 (s, 6 H), 2.81 (d, 2 H), 4.16 (m, 4 H), 5.40 (d, 1 H); IR (film) cm^{-1}: 2980, 1720, 1630, 1370, 1255. The absence of residual hexachloroethane was confirmed by ^{13}C NMR spectroscopy.

15. The phosphonate should be stored at 0°C. Under these conditions the purified product is stable for at least several months.

3. Discussion

This procedure is a modification of the previously-published procedure by Boeckman and Thomas.[6] The acetone diketene adduct serves as a versatile, activated β-keto ester equivalent.[2,3,7] Conversion of this material to the phosphonate by the procedure described above affords an even more versatile synthon which is useful for the preparation of protected analogues of the Nazarov reagents by means of a Wadsworth-Emmons olefination.[8,9]

$$(EtO)_2P(O)CH_2\text{-dioxinone} \xrightarrow[\text{THF, RCHO}]{\text{LDA}} R\text{-CH=CH-dioxinone}$$

The title phosphonate and related substances undergo thermal decomposition to β-acyl ketenes at temperatures in excess of 50°C.[10] Thus thermolysis in the presence of alcohols, amines, α-hydroxy esters, and α-amino esters affords the corresponding β-keto esters and amides; the latter two classes can be cyclized upon subsequent base treatment to unsaturated tetronic and tetramic acids and the related phosphonate reagents.[6,11]

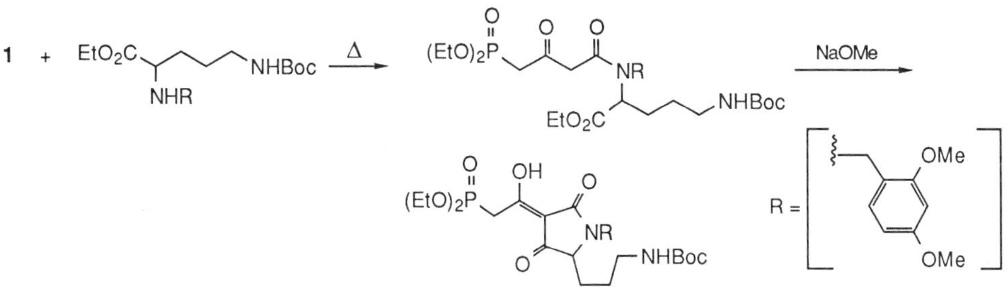

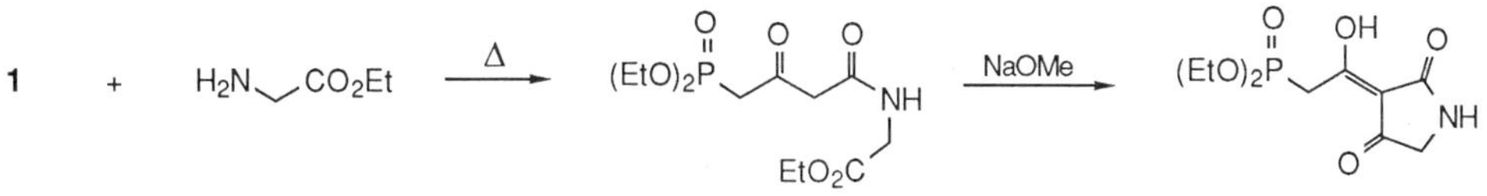

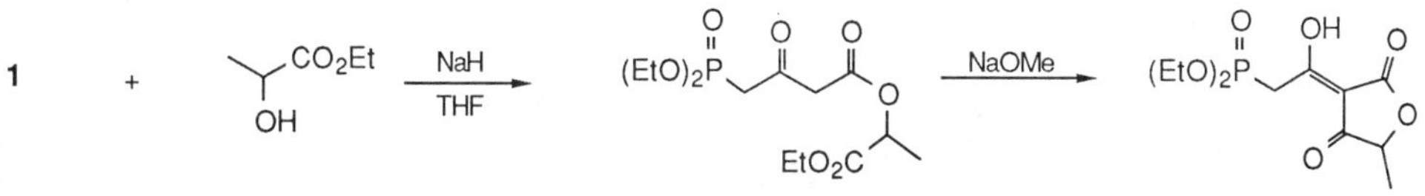

For sensitive amino acids prone to thermal dimerization to the related diketopiperazines, the reaction can be conducted in refluxing tetrahydrofuran solution in the presence of p-toluene- or camphorsulfonic acid as catalyst. Where possible the non acid-catalyzed thermal procedure is preferred since it generally provides cleaner products in higher yields.

The resulting tetramic and tetronic acid phosphonate reagents undergo the Wadsworth-Emmons olefination[9] with a variety of aldehydes to afford (E)-α,β-unsaturated and diene acyl tetramic and tetronic acids in good to excellent yields upon treatment with potassium tert-butoxide (2 equiv) in tetrahydrofuran. For readily enolizable substrates use of the N-protected systems is generally required. The following compounds have been prepared, in the indicated yields, in this manner:

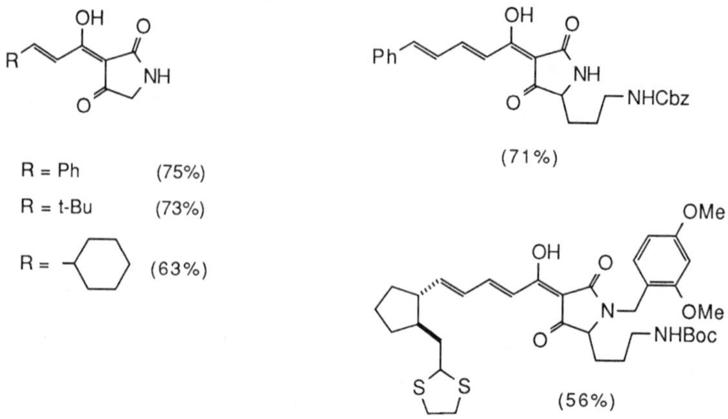

R = Ph (75%)
R = t-Bu (73%)
R = cyclohexyl (63%)

(71%)

(56%)

Two alternative methods for the preparation of phosphorus-activated tetramic acid reagents have recently been described.[12] These reagents have served to provide a workable solution to the problem of construction of the dienoyl tetramic acid unit required for the synthesis of tirandamycin-A.[13-16]

1. Department of Chemistry, University of Rochester, Rochester, NY 14627.
2. Carroll, M. F.; Bader, A. R. *J. Am. Chem. Soc.* **1953**, *75*, 5400.
3. (a) Gronowitz, S.; Hornfeldt, A. B.; Pettersson, K. *Synth. Commun.* **1973**, *3*, 213; (b) Kattenberg, J.; de Waard, E. R.; Huisman, H. O. *Tetrahedron* **1973**, *29*, 4149; (c) Smith, A. B., III; Scarborough, Jr., R. M. *Tetrahedron Lett.* **1978**, 4193.
4. Winkle, M. R.; Lansinger, J. M.; Ronald, R. C. *J Chem. Soc., Chem. Commun.* **1980**, 87.
5. Still, W. C.; Kahn, M. ; Mitra, A. *J. Org. Chem.* **1978**, *43*, 2923.
6. Boeckman, Jr., R. K.; Thomas, A. J. *J. Org. Chem.* **1982**, *47*, 2823.
7. Thomas, A. J. Ph.D. Dissertation, Wayne State University, 1983.

8. Nazarov, I. N.; Zav'yalov, S. I. *Zh. Obshch. Khim.* **1953**, *23*, 1703; *Chem. Abstr.* **1954**, *48*, 13667h.

9. Wadsworth, Jr., W. S. *Org. React.* **1977**, *25*, 73.

10. Hyatt, J. A.; Feldman, P. L.; Clemens, R. J. *J. Org. Chem.* **1984**, *49*, 5105.

11. Boeckman, Jr., R. K.; Moon, K. G. unpublished results, 1985.

12. (a) DeShong, P.; Lowmaster, N. E.; Baralt, O. *J. Org. Chem.* **1983**, *48*, 1149; (b) Schlessinger, R. H.; Bebernitz, G. R. *J. Org. Chem.* **1985**, *50*, 1344.

13. Schlessinger, R. H.; Bebernitz, G. R.; Lin P.; Poss, A. J. *J. Am. Chem. Soc.* **1985**, *107*, 1777.

14. DeShong, P.; Ramesh, S.; Elango, V.; Perez, J. J. *J. Am. Chem. Soc.* **1985**, *107*, 5219.

15. Boeckman, Jr., R. K.; Starrett, J. E.; Nickell, D. G.; Sum, P.-E. *J. Am. Chem. Soc.* **1986**, *108*, 5549.

16. Neukom, C.; Richardson, D. P.; Myerson, J. H.; Bartlett, P. A. *J. Am. Chem. Soc.* **1986**, *108*, 5559.

Appendix

Chemical Abstracts Nomenclature (Collective Index Number); (Registry Number)

6-Diethylphosphonomethyl-2,2-dimethyl-1,3-dioxen-4-one: Phosphonic acid, [(2,2-dimethyl-4-oxo-4H-1,3-dioxin-6-yl)methyl]-, diethyl ester (11); (81956-28-7)

6-Chloromethyl-2,2-dimethyl-1,3-dioxen-4-one: 4H-1,3-Dioxin-4-one, 6-(chloromethyl)-2,2-dimethyl- (11); (81956-31-2)

2,2,6-Trimethyl-1,3-dioxen-4-one: m-Dioxin-4-one, 2,2,6-trimethyl- (8,9); (5394-63-8)

Hexachloroethane: Ethane, hexachloro- (8,9); (67-72-1)

Diethyl phosphite: Phosphonic, acid, diethyl ester (8,9); (762-04-9)

(±)-trans-2-(PHENYLSULFONYL)-3-PHENYLOXAZIRIDINE

(Oxaziridine, 3-phenyl-2-(phenylsulfonyl)-)

A. PhCHO + $PhSO_2NH_2$ ⟶ $PhSO_2N{=}CHPh$

B. $PhSO_2N{=}CHPh$ $\xrightarrow[BnEt_3N^+\ Cl^-/NaHCO_3]{MCPBA}$ [oxaziridine: $PhSO_2$–N–C(Ph)(H) with O bridging N and C]

Submitted by Lal C. Vishwakarma, Orum D. Stringer, and Franklin A. Davis.[1]
Checked by James Pribish and Edwin Vedejs.

1. Procedure

A. N-Benzylidenebenzenesulfonamide. A 3-L, one-necked, round-bottomed flask is equipped with a mechanical stirrer (Note 1), Dean-Stark water separator (Note 2), double-walled condenser attached to an argon gas inlet and outlet needle connectors through a mineral oil bubbler. Into the flask are placed 150 g of 5 Å powdered molecular sieves (Note 3), 2.0 g of Amberlyst 15 ion exchange resin (Note 4), 157 g of benzenesulfonamide (Note 5), 1650 mL of dry toluene and 107.5 g (1.014 mol) of freshly-distilled benzaldehyde (Note 5). The reaction mixture is stirred and heated at reflux under an argon atmosphere. Water which separates during the reaction is periodically removed and refluxing is continued until water separation ceases (approximately 16 hr). The reaction mixture is cooled to room temperature without stirring and the insoluble materials are filtered through a 500-mL capacity sintered-glass funnel of medium porosity. The residue in the filter funnel is washed thoroughly with another 700 mL of toluene in three portions. The collected filtrate is concentrated with a rotary evaporator to give a thick yellow oily residue which usually solidifies on standing. The residue is triturated with 800 mL of distilled pentane and the solid is broken into a powder with the aid of a flat-ended glass rod. The solid is separated by filtration through a 500-mL sintered glass funnel of medium porosity, washed with distilled pentane (2 x 100 mL) and air dried. The yield is 212 g (87%); mp 76-80°C (Note 6).

Although of sufficient purity for the next step, the sulfonimine can be further purified by recrystallization. In a 2-L Erlenmeyer flask containing 150 mL of ethyl acetate is dissolved, with warming, 212 g of the crude sulfonimine. After the mixture is cooled to room temperature, about 400 mL of pentane is added and the solution is allowed to stand at room temperature for 2-3 hr. The colorless crystalline product is collected by filtration, washed with 100 mL of pentane and air dried to give 191.5 g (78%), mp 78-80°C. The washings and filtrate are combined and the volume reduced by about one third using a rotary evaporator. A second crop of crystals, 20.2 g (8%), mp 75-79°C, was obtained on standing for several hours.

B. (±)-trans-2-(Phenylsulfonyl)-3-phenyloxaziridine. A 5-L, three-necked flask is equipped with a mechanical stirrer and a 500-mL pressure-equalizing addition funnel. Into the flask are placed 500 mL of saturated aqueous sodium bicarbonate solution, 12.5 g (0.055 mol) of benzyltriethylammonium chloride (BTEAC) and 122.5 g (0.50 mol) of N-benzylidenebenzenesulfonamide dissolved in 380 mL of chloroform (Note 7). The reaction mixture is stirred vigorously at 0-5°C in an ice bath while a solution of 111.6 g (0.55 mol) of 85% m-chloroperoxybenzoic acid (MCPBA) dissolved in 1000 mL of chloroform is added dropwise. After the addition of the peracid, which takes about 1 hr, the reaction mixture is stirred for an additional hour at this temperature. A 3-L separatory funnel is used to separate the chloroform solution and wash it successively with 600 mL of cold water, 600 mL of aqueous 10% sodium sulfite, water (2 x 600 mL) and 250 mL of a saturated sodium chloride solution (Note 8). After the chloroform solution is dried over anhydrous potassium carbonate for 2 hr (Note 9), it is filtered and solvent is removed with a rotary evaporator, keeping the water bath temperature below 40°C. The resulting white solid residue is washed with a small portion of pentane, dissolved in a minimum of ethyl acetate (about 700-800 mL) without heating, and filtered through fluted filter paper; 400 mL of pentane is added to the filtrate. After the white crystalline oxaziridine is cooled in the refrigerator overnight, it is separated by filtration, transferred to a 500-mL Erlenmeyer flask, washed with 200 mL of pentane, filtered and air dried for 1 hr. The yield is 83.5 g; mp 92-94°C. The mother liquor is reduced to about 300 mL and cooled in the refrigerator to give 36.6 g of a light yellow solid; mp 87-90°C. This second crop is placed in a 250-mL Erlenmeyer flask and triturated with 50 mL of anhydrous ether followed by the addition of 60 mL of pentane. The oxaziridine is isolated by filtration to give 31.1 g; mp 94-95°C (dec) (Note 10, 11). The combined yield is 114.6 g (88%).

The 2-sulfonyloxaziridine can be stored in a brown bottle in the refrigerator. Storage at room temperature is potentially hazardous (Note 12).

2. Notes

1. A Teflon-coated, heavy duty, oval-shaped spin bar was used by the submitters for efficient stirring.

2. A Dean-Stark water separator equipped with a Teflon stopcock for water removal was used.

3. Linde powdered 5 Å molecular sieves were used as obtained from the supplier.

4. Amberlyst 15 ion-exchange resin is a strongly acidic, macroreticular resin purchased from Aldrich Chemical Company, Inc. The reaction fails in the absence of the acid catalysts.

5. Benzenesulfonamide, m-chloroperbenzoic acid and benzaldehyde were obtained from the Aldrich Chemical Company, Inc. and, with the exception of the latter, used without additional purification.

6. The ^1H NMR spectrum of N-benzylidenebenzenesulfonamide is as follows: ($CDCl_3$) δ: 7.6 (m, 6 H), 8.0 (m, 4 H) and 9.05 (s, 1 H).

7. Analytical reagent grade chloroform, Fisher Scientific Company, was used as obtained.

8. It is necessary to wash with a 10% $NaHCO_3$ solution, before the sodium chloride wash, if a large excess of m-chloroperoxybenzoic acid is used.

9. If the solution is dried for long times over potassium carbonate, decomposition of the oxaziridine sometimes occurs.

10. The ^1H NMR spectrum of trans-2-(phenylsulfonyl)-3-phenyloxaziridine is as follows: ($CDCl_3$) δ: 5.5 (s, 1 H), 7.4 (s, 5 H) and 7.6-7.8 (m, 3 H) and 8.05 (br d, 2 H, J = 7.1).

11. Careful recrystallization from ethyl acetate (saturated solution at 25°C; cool to -20°C) gave colorless crystals, ca. 20% recovery, mp 95-95.5°C.

12. Exothermic decomposition of a 500-g quantity after 2 weeks storage at room temperature is reported by Dr. G. C. Crockett, the Aldrich Chemical Company, Inc. Sufficient force was generated to shatter the container and char the oxaziridine.

3. Discussion

This procedure is representative of a general procedure for the synthesis of trans-2-sulfonyloxaziridines previously reported on a small scale (Table I).[2a] trans-2-(Phenylsulfonyl)-3-(p-nitrophenyl)oxaziridine was prepared on a 0.16-molar scale in greater than 85% yield. The Baeyer-Villiger-type oxidation of the sulfonimine affords only the trans-oxaziridine. The synthesis of the sulfonimine ($PhSO_2N{=}CHPh$) directly from the sulfonamide and aromatic aldehyde is described here. This modification avoids use of the intermediate diethyl acetal used in earlier preparations of these compounds.[2]

TABLE I

PREPARATION OF 2-SULFONYLOXAZIRIDINES[2a,b]

RSO_2N—CHAr (O bridging N and C)	% Yield	mp (°C, dec.)
R=Ph, Ar=3-NO_2Ph	83	113-5
R=Ph, Ar=4-NO_2Ph	80	134-6
R=Me, Ar=Ph	85	59-61
R=$PhCH_2$, Ar=Ph	90	118-9

2-Sulfonyloxaziridines are useful aprotic and neutral oxidizing reagents which, in general, afford greater selectivity for oxidations than do peracids. 2-Sulfonyloxaziridines have been employed in the oxidation of sulfides to sulfoxides,[3] disulfides to thiosulfinates,[3] selenides to selenoxides,[4] thiols to sulfenic acids (RSOH),[5] organometallic reagents to alcohols and phenols,[6] ketone and ester enolates to α-hydroxy carbonyl compounds,[7] in the epoxidation of alkenes,[8] and in the conversion of chiral amide enolates to optically active α-hydroxy carboxylic acids (93-99% ee).[9,10] These reagents can be used in the study of reactive oxidation intermediates and for mechanistic studies of oxygen-transfer reactions because of the ease with which the course of the oxidation can be monitored by proton NMR.

Oxygen-Transfer Reactions of 2-Sulfonyloxaziridines

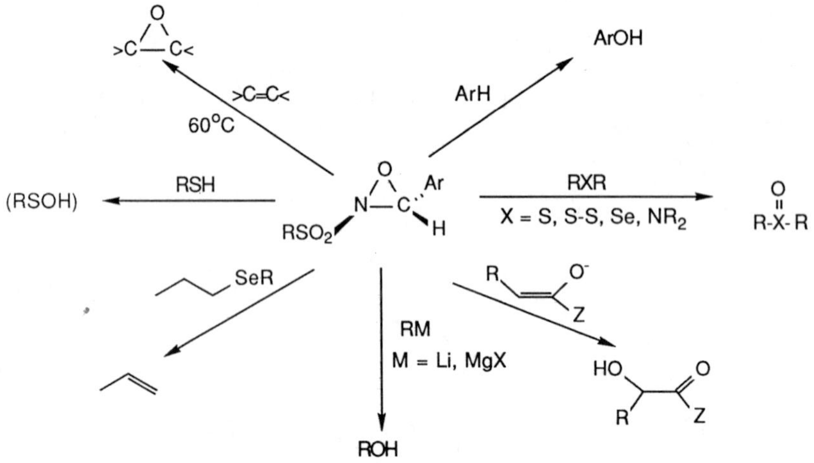

Oxidation of chiral sulfonimines ($R^{*}SO_2N{=}CHAr$)[11] and chiral sulfamylimines ($R^{*}RNSO_2N{=}CHAr$)[12] affords optically active 2-sulfonyloxaziridines and 2-sulfamyloxaziridines, respectively. These chiral, oxidizing reagents have been used in the asymmetric oxidation of sulfides to sulfoxides (15-68% ee),[11-13] selenides to selenoxides (8-9% ee),[14] enolates to α-hydroxycarbonyl compounds (8-37% ee)[15] and in the asymmetric epoxidation of alkenes (15-40% ee).[16]

1. Department of Chemistry, Drexel University, Philadelphia, PA 19104.

2. (a) Davis, F. A., Stringer, O. D. *J. Org. Chem.* **1982**, *47*, 1774; (b) Davis, F. A.; Lamendola, Jr., J.; Nadir, U.; Kluger, E. W.; Sedergran, T. C.; Panunto, T. W.; Billmers, R.; Jenkins, Jr., R.; Turchi, I. J.; Watson, W. H.; Chen, J. S.; Kimura, M. *J. Am. Chem. Soc.* **1980**, *102*, 2000.

3. Davis, F. A.; Jenkins, Jr., R.; Yocklovich, S. G. *Tetrahedron Lett.* **1978**, 5171; Davis, F. A.; Awad, S. B.; Jenkins, Jr., R. H.; Billmers, R. L.; Jenkins, L. A. *J. Org. Chem.* **1983**, *48*, 3071.

4. Davis, F. A.; Stringer, O. D.; Billmers, J. M. *Tetrahedron Lett.* **1983**, *24*, 1213.

5. Davis, F. A.; Rizvi, S. Q. A.; Ardecky, R.; Gosciniak, D. J.; Friedman, A. J.; Yocklovich, S. G. *J. Org. Chem.* **1980**, *45*, 1650; Davis, F. A.; Jenkins, Jr., R. H. *J. Am. Chem. Soc.* **1980**, *102*, 7967; Davis, F. A.; Billmers, R. H. *J. Am. Chem. Soc.* **1981**, *103*, 7016.

6. Davis, F. A.; Mancinelli, P. A.; Balasubramanian, K.; Nadir, U. K. *J. Am. Chem. Soc.* **1979**, *101*, 1044.

7. Davis, F. A.; Vishwakarma, L. C.; Billmers, J. M.; Finn, J. *J. Org. Chem.* **1984**, *49*, 3241.

8. Davis, F. A.; Abdul-Malik, N. F.; Awad, S. B.; Harakal, M. E. *Tetrahedron Lett.* **1981**, *22*, 917; Davis, F. A.; Abdul-Malik, N. F.; Jenkins, L. A. *J. Org. Chem.* **1983**, *48*, 5128.

9. Davis, F. A.; Vishwakarma, L. C. *Tetrahedron Lett.* **1985**, *26*, 3539.

10. Evans, D. A.; Morrissey, M. M.; Dorow, R. L. *J. Am. Chem. Soc.* **1985**, *107*, 4346.

11. Davis, F. A.; Jenkins, Jr., R. H.; Awad, S. B.; Stringer, O. D.; Watson, W. H.; Galloy, J. *J. Am. Chem. Soc.* **1982**, *104*, 5412.

12. Davis, F. A.; McCauley, Jr., J. P.; Harakal, M. E. *J. Org. Chem.* **1984**, *49*, 1465.

13. Davis, F. A.; Billmers, J. M. *J. Org. Chem.* **1983**, *48*, 2672.

14. Davis, F. A.; Stringer, O. D.; McCauley, Jr., J. P. *Tetrahedron* **1985**, *41*, 4747.

15. Boschelli, D.; Smith, III, A. B.; Stringer, O. D.; Jenkins, Jr., R. H; Davis, F. A. *Tetrahedron Lett.* **1981**, *22*, 4385.

16. Davis, F. A.; Harakal, M. E.; Awad, S. B. *J. Am. Chem. Soc.* **1983**, *105*, 3123.

Appendix

Chemical Abstracts Nomenclature (Collective Index Number); (Registry Number)

(±)-2-(Phenylsulfonyl)-3-phenyloxaziridine: Oxaziridine, 3-phenyl-2-(phenylsulfonyl)- (10); (63160-13-4)

N-Benzylidenebenzenesulfonamide: Benzenesulfonamide, N-benzylidene- (8); Benzenesulfonamide, N-(phenylmethylene)- (9); (13909-34-7)

TRISAMMONIUM GERANYL DIPHOSPHATE

(Diphosphoric acid, mono(3,7-dimethyl-2,6-octadienyl) ester (E)-, trisammonium salt)

A. geraniol (OH) → NCS, $(CH_3)_2S$ → geranyl chloride (Cl)

B. geranyl chloride (Cl) → 1. $(Bu_4N)_3P_2O_7H$; 2. Ion Exchange → $OP_2O_6(NH_4)_3$

Submitted by Andrew B. Woodside, Zheng Huang, and C. Dale Poulter.[1]
Checked by Pamela Seaton and James D. White.

1. Procedure

A. *Geranyl chloride.* To a flame-dried, 100-mL, three-necked, round-bottomed flask equipped with a magnetic stirrer, low temperature thermometer, rubber septum, and nitrogen inlet adapter, is added 1.47 g (11 mmol) of N-chlorosuccinimide (Note 1). The powder is dissolved in 45 mL of dry dichloromethane (Note 2), and the resulting solution is cooled to -30°C with a dry ice/acetonitrile bath. Freshly distilled dimethyl sulfide (0.87 mL, 0.74 g, 12 mmol) is added dropwise by syringe. The mixture is warmed to 0°C with an ice-water bath, maintained at that temperature for 5 min, and cooled to -40°C. To the resulting milky white suspension is added dropwise by syringe 1.54 g (10 mmol) of geraniol (Note 3) dissolved in 5 mL of dry dichloromethane. The suspension is warmed to 0°C with an ice-water bath and stirred for 2 hr. The ice bath is then removed, and the reaction mixture is allowed to warm to room temperature. Stirring is continued for an additional 15 min. The resulting clear, colorless solution is poured into a 250-mL separatory funnel and washed with 25 mL of saturated sodium chloride. The aqueous layer is washed with two 20-mL portions of pentane. The pentane extracts and an additional 20-mL portion of pentane are added to the methylene chloride extract. The resulting solution is washed twice with 10 mL of saturated sodium chloride and dried over magnesium sulfate. Solid material is removed by vacuum filtration through a fritted glass funnel, and most of the solvent is removed with a rotary evaporator at aspiratory pressure. The last traces of solvent are removed by pumping at high vacuum (0.2 mm) for 1.5 hr. The resulting pale yellow oil (1.61 g, 9.3 mmol, 93%) is used directly in the next step (Note 4).

B. Trisammonium geranyl diphosphate. To a flame-dried, 100-mL, round-bottomed flask equipped with a magnetic stirrer and a nitrogen inlet adapter is added 9.14 g (9.3 mmol) of tris(tetrabutylammonium) hydrogen pyrophosphate trihydrate (Note 5). The flocculant white solid is dissolved in 20 mL of dry acetonitrile (Note 6). To the resulting milky white suspension (Note 7) is added 0.83 g (4.8 mmol) of geranyl chloride. The mixure is allowed to stir at room temperature for 2 hr. Solvent is then removed with a rotary evaporator using a 40°C water bath. The pale yellow residue is dissolved in 3 mL of ion exchange buffer (Note 8), and the resulting clear solution is loaded onto a 4 x 15-cm column of Dowex AG 50W-X8 (100-200 mesh) cation exchange resin (ammonium form) (Note 9). The flask is washed twice with 5 mL of buffer and both washes are loaded onto the column before elution with 360 mL (two column volumes) of ion exchange buffer (Note 10). The eluant is collected in a 500-mL freeze-drying flask, frozen as described in Note 5, and lyophilized for 18-24 hr (Note 11) to yield 2.57 g of a white solid. The material is dissolved in 5 mL of 0.05 M ammonium bicarbonate, and the clear solution is transferred to a 50-mL centrifuge tube. Twenty milliliters of 1:1 (v/v) acetonitrile: isopropyl alcohol is added, and the contents are mixed thoroughly on a vortex mixer, during which time a white precipitate forms. The suspension is cleared by centrifugation for 5 min at 2000 rpm. The supernatant solution is removed with a pipette, the residue is suspended in 5 mL of 0.05 M ammonium bicarbonate, and the process is repeated. Three additional extractions are performed using 2 mL of 0.05 M ammonium bicarbonate and 8 mL of acetonitrile:isopropyl alcohol. The combined supernatant solutions (approximately 80 mL) are concentrated to approximately 5 mL with a rotary evaporator at 40°C (Note 12).

One half of the concentrated extract, dissolved in an equal volume of chromatography buffer (Note 13), is loaded onto a 5.5 x 18-cm cellulose flash column[2] (Note 14). The flask is rinsed with three 5-mL portions of chromatography buffer and each is loaded onto the column. The column is then eluted with 900 mL of chromatography buffer. After a 50-mL forerun, twenty-eight 30-mL fractions are collected, and every second fraction is analyzed by thin layer chromatography (Note 15). Fractions containing trisammonium geranyl diphosphate (typically 12-23) are pooled and concentrated to approximately 120 mL with a rotary evaporator at 40°C. The concentrate is transferred to a 250-mL freeze-drying flask and lyophilized for 18-24 hr as previously described in Note 5. The resulting flocculant white solid is collected and stored at -78°C. The cellulose chromatography is repeated to yield a total of 1.51-1.55 g (85-87%) of trisammonium geranyl diphosphate from geraniol (Note 16).

2. Notes

1. N-Chlorosuccinimide (from the Aldrich Chemical Company, Inc.) is recrystallized from benzene (*CAUTION: CARCINOGENIC*).

2. Methylene chloride is distilled from phosphorus pentoxide immediately before use.

3. Geraniol (from the Aldrich Chemical Company, Inc.) is distilled before use, bp 90-92°C at 3 mm.

4. The IR, ^{1}H NMR, and ^{13}C NMR spectra of this material are identical with those for distilled geranyl chloride (bp 49-51°C at 0.2 mm). Distillation on a small scale significantly reduces the yield, and there is no improvement in the yield of the phosphorylation reaction using distilled material. A synthesis of geranyl chloride was reported earlier in this series.[3] We find, however, that the procedure of Corey, Kim, and Takeda[4] is more convenient.

5. Disodium dihydrogen pyrophosphate (3.13 g, 14 mmol) (from Sigma Chemical Co.) is dissolved in 25 mL of deionized water containing 1 mL of concentrated ammonium hydroxide. The resulting clear solution is loaded onto a 2 x 30-cm column of Dowex AG 50W-X8 cation exchange resin (100-200 mesh, H^+) and eluted with deionized water. The first 150 mL of eluant is collected in a 250-mL freeze-drying flask. A magnetic stirring bar is added, and the solution is titrated to pH 7.3 by slow addition of tetrabutylammonium hydroxide (Aldrich Chemical Company, Inc.). The stirring bar is removed, and the flask is placed in a dry ice/propanol bath. The flask is spun slowly in a manner to uniformily freeze its contents to the walls. Water is removed by lyophilization for 24 hr. The resulting flocculant white solid (10.6 g, 83%) contains 3 to 4 waters of hydration and is used without further purification. The

material is extremely hygroscopic and can be stored in a desiccator over phosphorus pentoxide.

6. Reagent grade acetonitrile is distilled from phosphorus pentoxide immediately before use.

7. Clear solutions can be obtained by filtration. Residual water can be removed from freeze-dried salt by repeated evaporation of dry acetonitrile (rotary evaporator). This material is not noticeably more effective than the salt obtained after lyophilization.

8. Ammonium bicarbonate (2.0 g) is dissolved in 1 L of 2% (v/v) isopropyl alcohol/water. The resulting solution is 25 mM in ammonium bicarbonate.

9. The ammonium form of the resin is generated by placing 188 mL of Dowex AG 50W-X8 (100-200 mesh, H^+ form) in an 1-L feitted glass funnel and washing the material with four 200-mL portions of concentrated ammonium hydroxide. The resin is washed with 200-mL portions of deionized water until the pH of the filtrate drops to pH 7, then with two 200-mL portions of ion exchange buffer. The washed resin is suspended in 200 mL of buffer and slurry-packed into the column.

10. Dowex AG 50W-X8 (100-200 mesh) from BioRad has a capacity of 1.7 meq per mL of resin bed. This represents approximately a 10-fold excess of exchangeable ions in the resin over material loaded onto the column. However, the tetrabutylammonium cation has a lower affinity for the resin than the ammonium cation. To optimize the exchange, it is important to maintain a low concentration of ammonium ion in the exchange buffer, to elute the material slowly (less than 9 mL/min on the 4 x 15-cm column), and to elute with only two column volumes of exchange buffer; otherwise previously exchanged tetra-butylammonium cation will begin to elute from the column. Incomplete exchange

dramatically reduces the efficiency of the subsequent purification on cellulose. The efficiency of the exchange can be determined by ^{1}H NMR.

11. Trisammonium geranyl diphosphate will decompose if left under vacuum for extended periods. It is important to remove the sample from the freeze-drier within a few hours after water has been removed.

12. This material is stored at -20°C until chromatography on cellulose.

13. Chromatography buffer is prepared by dissolving 4.0 g of ammonium bicarbonate in 250 mL of deionized water and adding 500 mL of isopropyl alcohol and 250 mL of acetonitrile. The resulting solution is approximately 50 mM in ammonium bicarbonate.

14. Whatman CF11 fibrous cellulose powder is prepared for chromatography by the following procedure. Cellulose powder (1 L, dry volume) is mixed with 700 mL of deionied water in a 2-L beaker by vigorous stirring with a glass rod. The suspension is allowed to stand for 30 min, and the water is removed by decantation. The same procedure is followed as the cellulose is washed in succession with two 700-mL portions of 0.1 N hydrochloric acid, two 700-mL portions of deionized water, two 700-mL portions of 0.1 N sodium hydroxide, two 700-mL portions of deionized water, and two 700-mL portions of 1:1 (v/v) isopropyl alcohol:water. The material is stored at 4°C in 1:1 isopropyl alcohol:water until used. The column is slurry-packed in 1:1 isopropyl alcohol:water and washed with 1.3 L (approximately three colunn volumes) of acetonitrile. The column is then washed with 1.3 L of 1:1 isopropyl alcohol:water and equilibrated with 1.3 L of chromatography buffer (Note 13).

15. E. Merck cellulose thin-layer chromatography plates (available from American Scientific Products) are developed with chromatography buffer (Note 13) and visualized with sulfosalicylic acid/ferric chloride spray.[5] The system consists of a solution of 1.0 g of sulfosalicylic acid (from Aldrich

Chemical Co., Inc.) dissolved in 100 mL of 3:2 (v/v) ethanol:water and a solution of 0.20 g of ferric chloride in 100 mL of 4:1 (v/v) ethanol:water. Plates are first sprayed with the sulfosalicylic acid solution (thoroughly wetted but not dripping) and allowed to air dry. The ferric chloride solution is lightly sprayed onto the plates. Phosphate-containing compounds appear as white spots on a pink background. A second light spraying with ferric chloride may be necessary to make the spots pronounced. It is *important* to prepare both sprays freshly. Their shelf life is only about 6 hr. Under these conditions the trisammonium geranyl diphosphate has an R_f of 0.35. Residual tetrabutylammonium salt moves with the solvent front, and ammonium inorganic pyrophosphate remains at the origin.

16. This material migrates as a single spot on the cellulose thin-layer system and has no extraneous peaks in the ^1H, ^{13}C, and ^{31}P NMR spectra. The IR and NMR spectral properties of trisammonium geranyl diphosphate are as follows: IR (KBr) cm^{-1}: 3100-3500 (br), 2800,2990 (br), 1650, 1450, 1400, 1200, 1120, 1080, 1015, and 900; ^1H NMR (300 MHz, D_2O/ND_4OD) δ: 1.62 (s, 3 H, methy), 1.68 (s, 3 H, methyl) 1.72 (s, 3 H, methyl), 2.11 (m, 4 H, CH_2 at C_4 and C_5), 4.47 (t, 2 H, $J_{^1H,^1H}$ = 6.5, $J_{^1H,^{31}P}$ = 6.5, CH_2 at C_1), 5.22 (broad, 1 H, $J_{^1H,^1H}$ = 6.5, H at C_6), and 5.47 (t, 1 H, $J_{^1H,^1H}$ = 6.5, H at C_2); ^{13}C NMR (75 MHz, D_2O/ND_4OD, ^1H decoupled): 18.28 (CH_3), 19.69 (CH_3), 27.55 (CH_3), 28.60 (CH_2), 41.55 (CH_2), 65.42 (CH_2, d, $J_{^{13}C,^{31}P}$ = 4.0), 122.85 (CH, d, $J_{^{13}C,^{31}P}$ = 7.5), 127.10 (CH), 136.68 (C), and 145.76 (C); ^{31}P NMR (32 MHz, D_2O/ND_4OD, ^1H decoupled): -11.23 (d, 1 P, $J_{^{31}P,^{31}P}$ = 20) and -9.10 (d, 1 P, P_2).

3. Discussion

Previous methods for the preparation of salts of geranyl diphosphate and other allylic isoprenoid diphosphates are based on condensation between the alcohol and inorganic phosphate by trichloroacetonitrile as originally reported by Cramer[6] and modified by Popjak.[7] The reaction generates a complex mixture of organic and inorganic polyphosphates which must be separated by chromatography. The desired diphosphate ester has been prepared on small scales in yields of up to 30%,[8] but in our experience, the yields of pure material obtained by this procedure are usually less than 10%.

The direct displacement reaction can be used to prepare many of the common diphosphate esters in the isoprene biosynthetic pathway,[9] including isopentenyl diphosphate.[10,11] The yields are typically 60-90% from the alcohol, and the absence of phosphate polymers found in the Cramer procedure simplifies the purification step. We have also used the displacement procedure to prepare radio-labeled material for biosynthetic studies.[12]

1. Department of Chemistry, University of Utah, Salt Lake City, UT 84112.

2. Still, W. C.; Kahn, M.; Mitra, A. *J. Org. Chem.* **1978**, *43*, 2923.
3. Calzada, J. G.; Hooz, J. *Org. Synth.* **1974**, *54*, 63.
4. Corey, E. J.; Kim, C. U.; Takeda, M. *Tetrahedron Lett.* **1972**, 4339.
5. "CRC Handbook of Chromatography", Gunter, Z; Sherma, J. Eds.; CRC Press: Cleveland, OH, 1972; Vol. 2, p. 143.
6. Cramer, F.; Rittersdorf, W.; Bohm, W. *Liebigs Ann. Chem.* **1962**, *654*, 180.
7. Popjak, G.; Cornforth, J. W.; Cornforth, R. H.; Rhyage, R.; Goodman, D. S. *J. Biol. Chem.* **1962**, *237*, 56.

8. Banthorpe, D. V.; Christou, P. N.; Pink, C. R.; Watson, D. G. *Phytochemistry* **1983**, *22*, 2465.

9. Dixit, V. M.; Laskovics, F. M.; Noall, W. I.; Poulter, C. D. *J. Org. Chem.* **1981**, *46*, 1967.

10. Davisson, V. J.; Woodside, A. B.; Poulter, C. D. *Methods Enzymol.* **1985**, *110*, 130-144.

11. Davisson, V. J.; Woodside, A. B.; Neal, T. R.; Stremler, K. E.; Muehlbacher, M.; Poulter, C. D. J. Org. Chem. **1986**, *51*, 4768-4779.

12. Davisson, V. J.; Zabriskie, T. M.; Poulter, C. D. *Bioorg. Chem.* **1986**, *14*, 46-54.

Appendix

Chemical Abstracts Nomenclature (Collective Index Number); (Registry Number)

Geranyl chloride: 2,6-Octadiene, 1-chloro-3,7-dimethyl-, (E)- (8,9); (5389-87-7)

N-Chlorosuccinimide: Succinimide, N-chloro- (8); 2,5-Pyrrolidinedione, 1-chloro- (9); (128-09-6)

Geraniol: 2,6-Octadien-1-ol, 3,7-dimethyl-, (E)- (9); (106-24-1)

Tris(tetrabutylammonium) hydrogen pyrophosphate: 1-Butanaminium, N,N,N-tributyl-, diphosphate (3:1) (10); (76947-02-9)

Disodium dihydrogen pyrophosphate: Pyrophosphoric acid, disodium salt (8); Diphosphoric acid, disodium salt (9); (7758-16-9)

Tetrabutylammonium hydroxide: Ammonium tetrabutyl-, hydroxide (8); 1-Butanaminium, N,N,N-tributyl-, hydroxide (9); (2052-49-5)

ETHYL α-(HYDROXYMETHYL)ACRYLATE

(2-Propenoic acid, 2-(hydroxymethyl)-, ethyl ester)

$$(EtO)_2\underset{\underset{O}{\|}}{P}CH_2CO_2Et \quad + \quad HCHO\ (aqueous) \xrightarrow[20\text{-}45^{\circ}C]{K_2CO_3,\ H_2O} CH_2{=}\underset{\underset{CH_2OH}{|}}{C}CO_2Et$$

Submitted by J. Villieras and M. Rambaud.[1]

Checked by Christina M. J. Fox and James D. White.

1. Procedure

A. Ethyl α-(hydroxymethyl)acrylate, (Note 1). A 1000-mL, four-necked, round-bottomed flask is fitted with a mechanical stirrer, 250-mL pressure-equalizing funnel, condenser, and thermometer. Paraformaldehyde (48 g, 1.6 mol), 1 N phosphoric acid (4 mL) and water (110 mL) are heated at 90°C for 1.5 hr to form a clear aqueous formaldehyde solution. This solution is cooled to room temperature. Triethyl phosphonoacetate (89.6 g, 0.4 mol) is added to the flask and the solution is stirred at room temperature at 1000 rpm. A solution of potassium carbonate (60.7 g, 0.44 mol) in water (60 mL) is added at room temperature (first slowly: 10 mL in 10 min) and then more rapidly (40 min). The temperature reaches 35-40°C and must be maintained at this level (with a water bath if necessary). Stirring is continued for 5 min at 40°C after the end of the addition; then the mixture (liquid-liquid heterogenous mixture) must be cooled rapidly to room temperature using an ice bath (Note 2) while diethyl ether (200 mL) and brine (150 mL) are added. After decantation, the mixture is extracted with ether (three 100-mL portions). The combined organic layers are washed with brine (two 100-mL portions) (Note 3) and dried over magnesium sulfate; the solvents are evaporated under reduced pressure and the remaining oil is distilled to give a fraction at 65-70°C (1 mm) which weighs 38.5-41.6 g (74-80%), n_D^{20} 1.4494. The hydroxy ester is of high purity as shown by GLC analysis (25 m silica capillary OV-1 column) and spectral data (Notes 4, 5).

2. Notes

1. All manipulations should be carried out in a well-ventilated hood. The preparation requires the use of formaldehyde solution, and gives rise to ethyl acrylate as a secondary product, the amount of which increases if the addition of the carbonate solution is too rapid and the temperature rises to 45°C.

A freshly-opened supply of paraformaldehyde purchased from Aldrich Chemical Company, Inc. was used by the checkers. The use of commercial formaldehyde solutions which now contain up to 15% methanol leads to the formation of several by-products which cannot be separated by distillation from the α-(hydroxymethyl)acrylate.

2. This experimental procedure must be followed carefully to avoid partial decomposition of ethyl α-(hydroxymethyl)acrylate. The reaction is stopped rapidly after the addition of the carbonate solution (5 min) to prevent formation of high molecular weight by-products which result from transesterification and Michael addition, both of which occur in the basic medium. However, about 25% of the product is lost. Addition of diethyl ether during cooling minimizes side reactions.

3. Treatment with brine allows total elimination of base in the organic layer and prevents any side reaction during the distillation.

4. The spectral properties of ethyl α-(hydroxymethyl)acrylate are as follows: ^{1}H NMR (CCl_4) δ: 4.20 (2 H, CH_2-OH); 5.80 and 6.15 (2 H, CH_2=); ^{13}C NMR ($CDCl_3$) δ: 60.9 (CH_2OH); 124.8 ($\underline{C}H_2$=C); 140.2 (CH_2=$\underline{C}$); 166.5 (COOEt).

5. α-(Bromomethyl)-, α-(chloromethyl)-, α-(iodomethyl)-, and α-(fluoromethyl)acrylates are easily obtained from the α-(hydroxymethyl)acrylate[2] as illustrated in the following procedure.

A 500-mL, four-necked, round-bottomed flask is fitted with a mechanical stirrer, 100-mL pressure-equalizing addition funnel, reflux condenser capped with a drying tube (silica gel) and a thermometer (-90°C to +60°C). The flask is charged with a stirred solution of ethyl α-(hydroxymethyl)acrylate (33.84 g, 0.26 mol) in dry ether (250 mL) at -10°C. Phosphorus tribromide (34 g, 11.5 mL, 0.12 mol) is added slowly (15 min). The temperature is allowed to rise to 20°C and stirring is continued for 3 hr. Water (150 mL) is added at -10°C and the mixture is extracted with technical-grade hexane (three 50-mL portions). The organic phase is washed twice with a saturated sodium chloride solution (50 mL) and dried over magnesium sulfate. The solvents are removed with a rotary evaporator under reduced pressure, and the remaining oil is distilled to give ethyl α-(bromomethyl)acrylate, bp 85-87°C (20 mm) which weighs 43.8 g (87%), n_D^{20} 1.4502. The ester is of high purity as shown by GLC analysis on a capillary OV-1 column, and spectral data.

The spectral properties of ethyl α-(bromomethyl)acrylate are as follows: ^{1}H NMR (CCl_4) δ: 4.15 (2 H, CH_2Br); 5.90 and 6.22 (2 H, H_2C=); ^{13}C NMR ($CDCl_3$) δ: 29.2 (CH_2Br), 126.5 ($\underline{C}H_2$=C); 137.8 (CH_2=$\underline{C}$); 164.5 (COOEt).

3. Discussion

Ethyl α-(hydroxymethyl)acrylate can be used for the synthesis of chloro and bromomethyl acrylates. The fluoro and iodo compounds have been prepared easily by halogen exchange from ethyl α-(bromomethyl)acrylate.[2]

The same procedure can be applied to the synthesis of diethyl α-(bromomethyl)vinylphosphonate.[3,4] The keto analogs can be obtained in the same way.[5]

The procedure described here is relatively new and gives improved overall yields of 60-67% for the preparation of ethyl α-(bromomethyl)acrylate in two stages from commercially available starting materials. Other more complex and less productive procedures have been described.[6]

Ethyl α-(bromomethyl)acrylate has been used extensively for the synthesis of α-methylene lactones from ketones and aldehydes,[7] and α-methylene lactams, which are known for their cytotoxic activity,[7-9] from imines.[8]

1. Laboratoire de Synthèse Organique Sélective, associé au C.N.R.S. n°475, Faculté des Sciences, 2, rue de la Houssiniere, F-44072 Nantes, France.
2. Villieras, J.; Rambaud, M. *Synthesis* **1982**, 924.
3. Knochel, P.; Normant, J. F. *Tetrahedron Lett.* **1984**, *25*, 1475.
4. Rambaud, M.; Del Vecchio, A.; Villieras, J. *Synth. Commun.* **1984**, *14*, 833.
5. Villieras, J.; Rambaud, M. *Synthesis* **1983**, 300.
6. Ferris, A. F. *J. Org. Chem.* **1955**, *20*, 780; Block, Jr., P. *Org. Synth., Collect. Vol. 5*, **1973**, 381-383; Holm, A.; Scheuer, P. J. *Tetrahedron Lett.* **1980**, *21*, 1125; Ramarajan, K.; Ramalingam, K.; O'Donnell, D. J.; Berlin, K. D. *Org. Synth.* **1983**, *61*, 56.

7. Öhler, E.; Reininger, K.; Schmidt, U. *Angew. Chem.* **1970**, *82*, 480; *Angew. Chem., Intern. Ed. Engl.* **1970**, *9*, 457.; Schlewer, G.; Stampf, J. L.; Benezra, C. *J. Med. Chem.* **1980**, *23*, 1031; Heindel, N. D.; Minatelli, J. A. *J. Pharm. Sci.* **1981**, *70*, 84; Lee, K.-H.; Ibuka, T.; Kim, S.-H.; Vestal, B. R.; Hall, I. H.; Huang, E. S. *J. Med. Chem.* **1975**, *18*, 812; Howie, G. A.; Stamos, I. K.; Cassady, J. M. *J. Med. Chem.* **1976**, *19*, 309; Rosowsky, A.; Papathanasopoulos, N.; Lazarus, H.; Foley, G. E.; Modest, E. J. *J. Med. Chem.* **1974**, *17*, 672; Marchand, B.; Benezra, C *J. Med. Chem.* **1982**, *25*, 650; Boldrini, G. P.; Savoia, D.; Tagliavini, E.; Trombini, C.; Umani-Ronchi, A. *J. Org. Chem.* **1983**, *48*, 4108.

8. Belaud, C.; Roussakis, C.; Letourneux, Y.; El Alami, N.; Villieras, J. *Synth. Commun.* **1985**, *15*, 1233.

9. Grieco, P. A. *Synthesis* **1975**, 67; Wege, P.M.; Clark, R. D.; Heathcock, C. H. *J. Org. Chem.* **1976**, *41*, 3144.

Appendix

Chemical Abstracts Nomenclature (Collective Index Number); (Registry Number)

Ethyl α-(hydroxymethyl)acrylate: 2-Propenoic acid, 2-(hydroxymethyl)-, ethyl ester (9); (10029-04-6)

Triethyl phosphonoacetate: Acetic acid, phosphono-, triethyl ester (8); Acetic acid, (diethoxyphosphinyl)-, ethyl ester (9); (867-13-0)

Ethyl α-(bromomethyl)acrylate: Acrylic acid, 2-(bromomethyl)-, ethyl ester (8); 2-Propenoic acid, 2-(bromomethyl)-, ethyl ester (9); (17435-72-2)

Unchecked Procedures

Accepted for checking during the period January 1, 1987

through September 1, 1987. An asterisk (*) indicates that

the procedure has been subsequently checked.

In accordance with a policy adopted by the Board of Editors, beginning with Volume 50 and further modified subsequently, procedures received by the Secretary and subsequently accepted for checking will be made available upon request to the Secretary, if the request is accompanied by a stamped, self-addressed envelope. (Most manuscripts require 54¢ postage).

Address requests to:

Professor Jeremiah P. Freeman
Organic Syntheses, Inc.
Department of Chemistry
University of Notre Dame
Notre Dame, Indiana 46556

It should be emphasized that the procedures which are being made available are unedited and have been reproduced just as they were first received from the submitters. There is no assurance that the procedures listed here will ultimately check in the form available, and some of them may be rejected for publication in *Organic Syntheses* during or after the checking process. For this reason, *Organic Syntheses* can provide no assurance whatsoever that the procedures will work as described and offers no comment as to what safety hazards may be involved. Consequently, more than usual caution should be employed in following the directions in the procedures.

Organic Syntheses welcomes, on a strictly voluntary basis, comments from persons who attempt to carry out the procedures. For this purpose, a Checker's Report form will be mailed out with each unchecked procedure ordered. Procedures which have been checked by or under the supervision of a member of the Board of Editors will continue to be published in the volumes of *Organic Syntheses*, as in the past. It is anticipated that many of the procedures in the list will be published (often in revised form) in *Organic Syntheses* in future volumes.

2426 Vicinal Dicarboxylation of an Alkene: cis-1-Methylcyclohexane-1,2-dicarboxylic Acid
J.-P. Deprés and A. E. Greene, LEDSS, Université Scientifique et Médicale de Grenoble, BP 68, F-38402 Saint Martin d'Heres Cedex, France

2436 Methyl 2-Chloro-2-cyclopropylideneacetate
T. Liese, F. Seyed-Mahdavi, and A. de Meijere, Inst. für Organische Chemie und Biochemie, Universität Hamburg, Martin-Luther-King-Platz 6, 2000 Hamburg 13, Germany

2437 1-Chloro-1-(trichloroethenyl)cyclopropane
T. Liese, F. Jaekel, and A. de Meijere, Inst. für Organische Chemie und Biochemie, Universität Hamburg, Martin-Luther-King-Platz 6, 2000 Hamburg 13, Germany

2448* Palladium(0)-Catalyzed Syn-1,4-Addition of Carboxylic Acids to Cyclopentadiene Monoepoxide: Cis-3-Acetoxy-5-Hydroxycyclopent-1-ene
D. R. Deardorff and D. C. Myles, Department of Chemistry, Occidental College, Los Angeles, CA 90041

2449 Methyl 7-Hydroxyhept-5-ynoate
G. Casy, J. W. Patterson and R. J. K. Taylor, School of Chemical Science, University of East Anglia, Norwich, NR4 7TJ, United Kingdom

2450* R-(+)-1,1'-Binaphthalene-2,2'-diol
L. K. Truesdale and D. L. Coffen, Research Department, Hoffmann-La Roche Inc., Nutley, NJ 07110

2452 Selective Paraffin Hydroxylations with Activated Peroxycarboxylic Acids: 7aH-cis-3a-Hydroxy Tetrahydroindene
H.-J. Schneider, W. Müller and N. Nguyen-Ba, Fachrichtung Organische Chemie der Universität des Saarlandes, D-6600 Saarbrücken 11, West Germany

2453* The Carroll Rearrangement: Synthesis of 5-Dodecen-2-one
S. R. Wilson and C. E. Augelli, Department of Chemistry, New York University, New York, NY 10003

2455* Preparation of 1,4-Di-O-Alkyl Threitols from Tartaric Acid: 1,4-Di-O-Benzyl-L-Threitol
E. A. Mash, K. A. Nelson, S. Van Deusen, and S. B. Hemperly, Department of Chemistry, The University of Arizona, Tucson, AZ 85721

2457 Enantioselective Oxidation of Sulfides: Synthesis of (-)-(S)-Methyl p-Tolyl Sulfoxide
S. H. Zhao, O. Samuel, and H. B. Kagan, Laboratoire de Synthese Asymetrique, Universite Paris-Sud, 91405 Orsay, France

CUMULATIVE AUTHOR INDEX

FOR VOLUMES 65 AND 66

This index comprises the names of contributors to Volume **65** and **66** only. For authors to previous volumes, see cumulative indices in Volume **64**, which covers Volumes **60** through **64**, and Volume **59**, which covers Volumes **55** through **59**, and either indices in Collective Volumes I through V or single volumes entitled *Organic Syntheses, Collective Volumes, I, II, III, IV, V, Cumulative Indices*, edited by R. L. Shriner and R. H. Shriner.

CUMULATIVE SUBJECT INDEX
FOR VOLUMES 65 AND 66

This index comprises subject matter for Volumes **65** and **66** only. For subjects in previous volumes, see the cumulative indices in Volume **64**, which covers Volumes **60** through **64**, and Volume **59**, which covers Volumes **55** through **59**, and either the indices in Collective Volumes I through V or the single volume entitled *Organic Syntheses, Collective Volumes I, II, III, IV, V, Cumulative Indices*, edited by R. L. Shriner and R. H. Shriner.

The index lists the names of compounds in two forms. The first is the name used commonly in procedures. The second is the systematic name according to **Chemical Abstracts** nomenclature, accompanied by its registry number in brackets. While the systematic name is indexed separately, it also accompanies the common name. Also included are general terms for classes of compounds, types of reactions, special apparatus, and unfamiliar methods.

Most chemicals used in the procedure will appear in the index as written in the text. There generally will be entries for all starting materials, reagents, intermediates, important by-products, and final products. Entries in capital letters indicate compounds, reactions, or methods appearing in the title of the preparation.

Acetic acid, chloro-, 5-methyl-2-(1-methyl-1-phenylethyl)cyclohexyl ester, [1R-(1α,2β,5α)]-, **65**, 203

Acetic acid, (diethoxyphosphinyl)-, ethyl ester; (867-13-0), **66**, 224

Acetic acid ethenyl ester, **65**, 135

Acetic acid, hydrazinoimino-, ethyl ester; (53085-26-0), **66**, 149

Acetic acid, trifluoro-, anhydride, **65**, 12

Acetic acid vinyl ester, **65**, 135

ACETONE TRIMETHYLSILYL ENOL ETHER: SILANE, (ISOPROPENYLOXY)TRIMETHYL-; SILANE, TRIMETHYL[(1-METHYLETHENYL)OXY]-; (1833-53-0), **65**, 1

Acetonitrile, purification, **66**, 101

Acetophenone; Ethanone, 1-phenyl-; (98-86-2), **65**, 6, 119

Acetophenone silyl enol ether: Silane, trimethyl[(1-phenylvinyl)oxy]-; Silane, trimethyl[(1-phenylethenyl)oxy]-; (13735-81-4), **65**, 12

4-ACETOXYAZETIDIN-2-ONE: 2-AZETIDINONE, 4-HYDROXY-ACETATE (ESTER): 2-AZETIDINONE, 4-(ACETYLOXY)-; (28562-53-0), **65**, 135

Acetylene; Ethyne; (74-86-2), **65**, 61

α,β-Acetylenic esters from β-chloroalkylidene malonates, **66**, 178

3-ACETYL-4-HYDROXY-5,5-DIMETHYLFURAN-2(5H)-ONE, **66**, 108, 110

Acrylonitrile; 2-Propenenitrile; (107-13-1), **65**, 236

Acyl chlorides, preparation, **66**, 116

Acylsilanes, preparation, **66**, 18

ADDITION OF TIN RADICALS TO TRIPLE BONDS, **66**, 75

Adipic acid monomethyl ester; (627-91-8), **66**, 116, 117, 120

Alane, (E)-1-decenyldiisobutyl-, **66**, 60, 61

(S)-Alanine (L-alanine); (56-41-7), **66**, 151-153, 159

Aldehydes, preparation from carboxylic acid, **66**, 124

Aldol reaction, **65**, 6, 12

Diels-Alder reaction of triethyl 1,2,4-triazine-3,5,6-tricarboxylate, **66**, 150

Diels-Alder reactions, **66**, 40

Diethylamine, **66**, 145

Diethyl aminomethylphosphonate: Phosphonic acid, (aminomethyl)-, diethyl ester; (50917-72-1), **65**, 119

DIETHYL N-BENZYLIDENEAMINOMETHYLPHOSPHONATE: PHOSPHONIC ACID, [[(PHENYLMETHYLENE)AMINO]METHYL]-, DIETHYL ESTER; (50917-73-2), **65**, 119

Diethyl 2-chloro-2-cyclopropylethene-1,1-dicarboxylate, **66**, 173, 174

Diethyl cyclopropylcarbonylmalonate; (7394-16-3), **66**, 175, 179

Diethyl cyclopropylmalonate, **66**, 173

Diethyl dioxosuccinate; (59743-08-7), **66**, 144, 149

Diethyl isocyanomethylphosphonate: Phosphonic acid, (isocyanomethyl)-, diethyl ester; (41003-94-5), **65**, 119

Diethyl malonate; (105-53-3), **66**, 175, 179

Diethyl oxalate: Oxalic acid, diethyl ester; Ethanedioic acid, diethyl ester; (95-92-1), **65**, 146

Diethyl phosphite; (762-04-9), **66**, 195, 197, 202

6-DIETHYLPHOSPHONOMETHYL-2,2-DIMETHYL-1,3-DIOXEN-4-ONE, **66**, 194-196, 202

Diethyl phthalimidomethylphosphonate: Phosphonic acid, (phthalimidomethyl)-, diethyl ester; Phosphonic acid, [(1,3-dihydro-1,3-dioxo-2H-isoindol-2-yl)-methyl]-, diethyl ester; (33512-26-4), **65**, 119

DIHYDROJASMONE, **65**, 26

1,2-Dihydroxybenzene; (120-80-9), **66**, 180, 184

Dihydroxytartaric acid disodium salt hydrate; (866-17-1), **66**, 144, 146, 149

Diisobutylaluminum hydride; (1191-15-7), **66**, 60, 62, 66, 186, 188, 193

ETHYL 4-CYCLOHEXYL-4-OXOBUTANOATE: CYCLOHEXANEBUTANOIC ACID, γ-OXO-, ETHYL ESTER; (54966-52-8), **65**, 17

ETHYL CYCLOPROPYLPROPIOLATE, **66**, 173, 174

ETHYL (E,Z)-2,4-DECADIENOATE; (3025-30-7), **66**, 22, 23, 25, 28

Ethyl 3,4-decadienoate; (36186-28-4), **66**, 22, 28

Ethylenediaminetetraacetic acid, tetrasodium salt: Glycine, N,N'-1,2-ethanediylbis[N-(carboxymethyl)]-, tetrasodium salt, trihydrate; (67401-50-7), **65**, 166

Ethyl ether, compd. with boron fluoride (BF_3) (1:1), **65**, 17

ETHYL α-(HYDROXYMETHYL)ACRYLATE; (10029-04-6), **66**, 220, 222, 224

Ethyl N-(2-methyl-3-nitrophenyl)formimidate, **65**, 146

Ethyl oxalamidrazonate; (53085-26-0), **66**, 143, 146, 149

Ethyl propiolate; (623-47-2), **66**, 29, 31, 36

Ethyl thioamidooxalate; (16982-21-1), **66**, 143, 145, 149

Ethyne, **65**, 61

Finkelstein reaction, **66**, 87

Flash chromatography, **66**, 135, 196

Florisil, **66** 197

Formaldehyde, **66**, 220

Formic acid, chloro-, ethyl ester; (541-41-3), **66**, 141

Formic acid, chloro-, methyl ester, **65**, 47

Formic acid, cyano-, ethyl ester; (623-49-4), **66**, 149

Geraniol; (106-24-1), **66**, 211, 214, 219

Geranyl chloride; (5389-87-7), **66**, 211, 212, 214, 219

GERANYL DIPHOSPHATE, TRISAMMONIUM SALT, **66**, 211-213

Glucopyranoside, methyl, α-D-, **65**, 243

2-Methyl-2-butene: 2-Butene, 2-methyl-; (513-35-9), **65**, 159

(3-Methyl-2-butenyl)propanedioic acid, dimethyl ester;
(43219-18-7), **66**, 75, 85

(3-Methyl-2-butenyl)(2-propynyl)propanedioic acid, dimethyl ester, **66**, 76

Methyl carbamate: Carbamic acid, methyl ester; (598-55-0), **65**, 159

METHYL 4-CHLORO-2-BUTYNOATE: 2-BUTYNOIC ACID, 4-CHLORO-, METHYL ESTER;
(41658-12-2), **65**, 47

Methyl chloroformate: Formic acid, chloro-, methyl ester;
Carbonochloridic acid, methyl ester; (79-22-1), **65**, 47

Methyl (E)-crotonate; (623-43-8), **66**, 38, 39, 41, 42

1-Methylcyclohexene: Cyclohexene, 1-methyl-; (591-49-1), **65**, 90

3-Methyl-2-cyclohexen-1-one; (1193-18-6), **66**, 37, 39, 42

Methyl N,N-dichlorocarbamate: Carbamic acid, dichloro-, methyl ester;
(16487-46-0), **65**, 159

Methyl 1,3-dimethyl-5-oxobicyclo[2.2.2]octane-2-carboxylate, **66**, 37

Methyl α-D-glucopyranoside: Glucopyranoside, methyl, α-D-;
α-D-glucopyranoside, methyl; (97-30-3), **65**, 243

4-METHYL-3-HEPTANONE, (S)-(+)-: 3-HEPTANONE, 4-METHYL-, (S)-;
(51532-30-0), **65**, 183

4-Methyl-3-heptanone SAMP-hydrazone, (S)-(+)-: 1-Pyrrolidinamine,
N-(1-ethyl-2-methylpentylidene)-2-(methoxymethyl)-,
[S-[R*,R*-(Z)]-; (69943-24-4), **65**, 183

Methyl 2-hydroxyisobutyrate; (2110-78-3), **66**, 110, 111, 114

Methyllithium: Lithium, methyl-; (917-54-4), **65**, 47, 140; **66**, 97, 100

Methyllithium, low halide, **66**, 53, 55

Methylmagnesium chloride, **66**, 1-3

[(Methyl)(methoxy)carbene]pentacarbonyl chromium(0): Chromium, pentacarbonyl(1-methoxyethylidene)-, (OC-6-21)-; (20540-69-6), **65**, 140

METHYL N-(2-METHYL-2-BUTENYL)CARBAMATE: CARBAMIC ACID, (2-METHYL-2-BUTENYL)-, METHYL ESTER; (86766-65-6), **65**, 159

5-Methyl-2-(1-methyl-1-phenylethyl)cyclohexanone, (2R,5R)-: Cyclohexanone, 5-methyl-2-(1-methyl-1-phenylethyl)-, (2R-trans)-; (57707-92-3), **65**, 203

5-Methyl-2-(1-methyl-1-phenylethyl)cyclohexanone, (2S,5R)-: Cyclohexanone, 5-methyl-2-(1-methyl-1-phenylethyl-, (2S-cis)-; (65337-06-6), **65**, 203

5-Methyl-2-(1-methyl-1-phenylethyl)cyclohexyl chloroacetate, (1R,2S,5R)-: Acetic acid, chloro-, 5-methyl-2-(1-methyl-1-phenylethyl)cyclohexyl ester, [1R-(1α,2β,5α)]-; (71804-27-8), **65**, 203

5-Methyl-2-[1-methyl-1-(phenylmethylthio)ethyl]cyclohexanone, cis- and trans-; **65**, 215

5-Methyl-2-(1-methyl-1-thioethyl)cyclohexanol, **65**, 215

N-Methylmorpholine, **66**, 133, 135

2-Methyl-3-nitroaniline: o-Toluidine, 3-nitro-; Benzeneamine, 2-methyl-3-nitro-; (603-83-8), **65**, 146

2-METHYL-4'-NITROBIPHENYL, **66**, 67, 68

(R)-METHYLOXIRANE; (15448-47-2), **66**, 160, 161, 163, 164, 172

METHYL 6-OXODECANOATE; (61820-00-6), **66**, 116, 117, 120, 123

METHYL 2-OXO-5,6,7,8-TETRAHYDRO-2H-1-BENZOPYRAN-3-CARBOXYLATE: 2H-1-BENZOPYRAN-3-CARBOXYLIC ACID, 5,6,7,8-TETRAHYDRO-2-OXO-, METHYL ESTER; (85531-80-2), **65**, 98

(E)-3-Methyl-3-penten-2-one, **66**, 11

(Z)-3-Methyl-3-penten-2-one, **66**, 11

2-Sulfonyloxaziridines, preparation, **66**, 207, 208

Sulfosalicylic acid spray for tlc plates, **66**, 216

Sulfur chloride, **65**, 159

Sulfur dichloride: Sulfur chloride; (10545-99-0), **65**, 159

Sulfur diimide, dicarboxy-, dimethyl ester, **65**, 159

Sulfuryl chloride isocyanate, **65**, 135

Swern oxidation, **6**, 15, 18

2,3,4,6-Tetra-O-acetyl-α-D-glucopyranosyl bromide: Glucopyranosyl bromide tetraacetate, α-D-; α-D-glucopyranosyl bromide, 2,3,4,6-tetraacetate; (572-09-8), **65**, 236

Tetrabutylammonium fluoride; (429-41-4), **66**, 110, 111, 115

Tetrabutylammonium hydroxide; (2052-49-5), **66**, 214, 219

Tetrafluoroboric acid-diethyl ether complex; (67969-82-8), **66**, 97, 98, 100, 107

Tetraisopropyl titanate: Isopropyl alcohol, titanium (4+) salt; 2-Propanol, titanium (4+) salt; (546-68-9), **65**, 230

Tetrakis(acetonitrile)palladium tetrafluoroborate; (21797-13-7), **66**, 52, 54, 59

Tetrakis(triphenylphosphine)palladium; (14221-01-3), **66**, 61, 62, 66, 68, 69, 74

4,5,4',5'-TETRAMETHOXY-1,1'-BIPHENYL-2,2'-DICARBOXALDEHYDE: [1,1'-BIPHENYL]-2,2'-DICARBOXALDEHYDE, 4,4',5,5'-TETRAMETHOXY-; (29237-14-7), **65**, 108

Tetramethyltin, **66**, 55